Declan A. Diver

A Plasma Formulary
for Physics, Technology and Astrophysics

Declan A. Diver

A Plasma Formulary for Physics, Technology and Astrophysics

WILEY-VCH

Berlin · Weinheim · New York · Chichester · Brisbane · Singapore · Toronto

Author:
Dr. Declan A. Diver, Department of Physics & Astronomy, University of Glasgow, U.K.
e-mail: diver@astro.gla.ac.uk

Cover:
Solar image from the NASA TRACE satellite. With kind permission of NASA (background). A plasma plume created by laser ablation of a solid surface. With kind permission of Dr. K.W.D. Ledingham, Department of Physics & Astronomy, University of Glasgow, UK (left). Atmospheric glow discharge between glass electrodes. With kind permission of Prof. W. Graham and Dr. P. Steen, Queen's University Belfast, UK (right).

1st edition

Library of Congress Card No: applied for

British Library Cataloguing-in-Publication Data: A catalogue record for this book is available from the British Library.

Die Deutsche Bibliothek – CIP Cataloguing-in-Publication-Data
A catalogue record for this publication is available from Die Deutsche Bibliothek

© WILEY-VCH Verlag Berlin GmbH, Berlin (Federal Republic of Germany), 2001

ISBN 3-527-40294-2

Printed on non-acid paper.

Printing: StraussOffsetdruck GmbH, Mörlenbach
Bookbinding: Wilhelm Osswald & Co., Neustadt (Weinstraße)

Printed in the Federal Republic of Germany.

WILEY-VCH Verlag Berlin GmbH
Bühringstrasse 10
D-13086 Berlin

To Anne, Caitlin and Ronan

Contents

List of Tables

Preface

Plasma physics has matured rapidly as a scientific and technological discipline with a vast span of relevant application in many different fields. As a consequence, no single textbook is able to address all aspects of plasma physics relevant to such a burgeoning community.

With this reference text I have attempted to bridge the gap between the excellent variety of traditional, broadly-based plasma books, and more specialist, device-oriented reference texts. David L Book's NRL Plasma Formulary was an inspiration, as too was André Anders' Formulary for Plasma Physics; however, I believe that this book offers a different perspective which makes it complementary to existing handbooks. I have tried to give the reader an overview of the key aspects of plasma physics without being too specialist in any particular area. Since this book is not a textbook, there is more room for not just contemporary findings, but also many traditional established results from the 1950's and 60's that are not often found in modern texts, and which are once more becoming important as imperfectly ionised and bounded plasmas enjoy a resurgence in relevance.

The diverse nature of the plasma science community is matched by a confusing miscellany of physical units. Throughout this handbook, all formulae are quoted in both SI and cgs-Guassian units where it is relevant. I hope this will maximise this book's practicality and utility, and perhaps even assist the whole community in the smooth transition to using SI units only....

It has been a guiding principle to reference the source (or sources) of any formula quoted in this book, together with whatever caveats or restrictions that apply to its use. Where practical I have referenced the original articles, subject to the important constraint that verifiable sources are accessible to the general reader. Please accept my apologies in advance for any misquotes or omissions, and please do let me know about them. As for the formulae themselves, I am indebted to Prof E W Laing for his patient and exacting examination of the manuscript, which led to the elimination of a very large number of errors. Thanks are also due to my colleagues Brendan Dowds, Hugh Potts, Richard Barrett, Graham Woan, Norman Gray and Graeme Stewart, for answering endless questions on $\text{\LaTeX} \, 2_\varepsilon$ formatting and graphics, and pointing out still more howlers in the ith iterate of the book. Despite all this invaluable and talented assistance, I have no doubt that there remain, lurking in dark corners of the text, or even brazenly displayed in large, open areas, errors in physics and formatting. I have no excuse; please let me know, and I shall make good these mistakes.

I am also grateful to Prof Ken Ledingham for letting me use his wonderful image of a laser-produced plasma plume; likewise, to Prof Bill Graham for the beautiful high-pressure discharge picture.

It is appropriate to acknowledge the kind support offered by David Hughes in guiding me initially on this project, and latterly Vera Dederichs for patiently enduring one delay after another in its prosecution. Thanks are also due to Prof A E Roy for wise advice at the outset. Finally, I am grateful to my Institute for granting me the sabbatical leave which was instrumental in allowing me to complete the book.

DECLAN ANDREW DIVER

Glasgow, July 2001

1
Basic Physical Data

1.1 BASIC PHYSICAL UNITS

1.1.1 SI Units

Table 1.1: Fundamental and supplementary SI units

QUANTITY	UNIT	ABBREVIATION
Fundamental Units		
mass	kilogram	kg
length	metre	m
time	second	s
temperature	Kelvin	K
electrical current	ampere	A
luminous intensity	candela	cd
amount of substance	mole	mol
plane angle	radian	rad
solid angle	steradian	sr
Selected derived units		
frequency	hertz	Hz
force	newton	N
energy	joule	J
power	watt	W
electrical charge	coulomb	C
electric potential	volt	V
electrical resistance	ohm	Ω
capacitance	farad	F
inductance	henry	H
magnetic flux	weber	Wb
magnetic flux density	tesla	T

Table 1.2: Standard prefixes for SI units

PREFIX	SYMBOL	FACTOR	PREFIX	SYMBOL	FACTOR
yotta	Y	10^{24}	deci	d	10^{-1}
zetta	Z	10^{21}	centi	c	10^{-2}
exa	E	10^{18}	milli	m	10^{-3}
peta	P	10^{15}	micro	μ	10^{-6}

continued on next page

Table 1.2: *continued*

PREFIX	SYMBOL	FACTOR	PREFIX	SYMBOL	FACTOR
tera	T	10^{12}	nano	n	10^{-9}
giga	G	10^{9}	pico	p	10^{-12}
mega	M	10^{6}	femto	f	10^{-15}
kilo	k	10^{3}	atto	a	10^{-18}
hecto	h	10^{2}	zepto	z	10^{-21}
deca	da	10^{1}	yacto	y	10^{-24}

1.1.2 cgs-Gaussian Units

For a useful overview of non-SI units see [15].

Table 1.3: Comparison of SI and cgs units

QUANTITY	UNIT	ABBREV.	SI EQUIVALENT
length	centimetre	cm	10^{-2}m
mass	gramme	g	10^{-3}kg
time	second	s	1s
force	dyne	dyn	10^{-5}N
energy	erg	erg	10^{-7}J
power	erg per second	$\mathrm{erg\,s^{-1}}$	10^{-7}W
electrical charge	statcoulomb	statcoul	$(3 \times 10^{9})^{-1}$C
current	statamp	statamp	$(3 \times 10^{9})^{-1}$A
electric potential	statvolt	statvolt	300V
magnetic flux density	gauss	G	10^{-4}T

1.2 MAXWELL'S ELECTROMAGNETIC EQUATIONS

Table 1.4: Maxwell's equations

	SI	cgs-Gaussian	
$\nabla \times \boldsymbol{E}$	$= -\dfrac{\partial \boldsymbol{B}}{\partial t}$	$= -\dfrac{1}{c}\dfrac{\partial \boldsymbol{B}}{\partial t}$	Faraday's law
$\nabla \times \boldsymbol{H}$	$= \dfrac{\partial \boldsymbol{D}}{\partial t} + \boldsymbol{J}$	$= \dfrac{1}{c}\dfrac{\partial \boldsymbol{D}}{\partial t} + \dfrac{4\pi}{c}\boldsymbol{J}$	Ampere's law

continued on next page

Table 1.4: *continued*

	SI	cgs-Gaussian	
$\nabla \cdot \boldsymbol{D}$	$= \rho_c$	$= 4\pi\rho_c$	Poisson equation
$\nabla \cdot \boldsymbol{B}$	$= 0$	$= 0$	
$\boldsymbol{D}$	$= \epsilon_r \epsilon_0 \boldsymbol{E}$	$= \epsilon_r \boldsymbol{E}$	
$\boldsymbol{B}$	$= \mu_r \mu_0 \boldsymbol{H}$	$= \mu_r \boldsymbol{H}$	

Boundary Conditions The boundary conditions at an interface for Maxwell's electromagnetic equations are that the tangential component of $\boldsymbol{E}$, and the normal component of $\boldsymbol{B}$, must each be continuous, where normal means parallel to the local normal vector to the interface, and tangential means in the plane perpendicular to the local normal.

1.3   SPECIAL RELATIVITY

Assume standard inertial frames S and S', with respective cartesian coordinates (x, y, z), (x', y', z') aligned such that the origins O, O' are co-incident at time $t = t' = 0$, with S' moving with velocity $\boldsymbol{v}$ with respect to S. Subscript $\parallel$ will denote the direction of this mutual motion, and subscript $\perp$ denotes the orthogonal plane. The Lorentz transformations of various physically significant quantities are given in the following table [61]:

Table 1.5: Lorentz transformations

QUANTITY	TRANSFORMATION
space-time:	$\boldsymbol{r} = \gamma_v(\boldsymbol{r}'_{\parallel} + \boldsymbol{v}t') + \boldsymbol{r}'_{\perp}$
	$t = \gamma_v(t + v r'_{\parallel}/c^2)$
invariant:	$r^2 - c^2 t^2$
velocity:	$\boldsymbol{u} = (\boldsymbol{u}'_{\parallel} + \boldsymbol{v} + \boldsymbol{u}'_{\perp}/\gamma_v)/(1 + u'_{\parallel} v/c^2)$
momentum-mass:	$\boldsymbol{p} = \gamma_v(\boldsymbol{p}'_{\parallel} + m'\boldsymbol{v}) + \boldsymbol{p}'_{\perp}$
	$mc = \gamma_v(m'c + v p'_{\parallel}/c)$
invariant:	$p^2 - m^2 c^2$
current & charge densities:	$\boldsymbol{J} = \gamma_v(\boldsymbol{J}'_{\parallel} + \boldsymbol{v}\rho_c) + \boldsymbol{J}'_{\perp}$

continued on next page

Table 1.5: *continued*

QUANTITY	TRANSFORMATION
invariant:	$\rho_c = \gamma_v(\rho_c' + vJ_\parallel'/c^2)$ $J^2 - c^2\rho_c^2$
electric & magnetic fields:	$\boldsymbol{E} = \boldsymbol{E}_\parallel' + \gamma_v(\boldsymbol{E}_\perp - \boldsymbol{v} \times \boldsymbol{B}')$ $\boldsymbol{B} = \boldsymbol{B}_\parallel' + \gamma_v(\boldsymbol{B}_\perp' + \boldsymbol{v} \times \boldsymbol{E}'/c^2)$

1.4 PHYSICAL CONSTANTS

The values of the constants quoted here are the 1998 CODATA recommended values [66], reproduced with permission.

Table 1.6: Values of physical constants

QUANTITY	SYMBOL	VALUE	UNITS
speed of light in vacuum	c	299 792 458	m s^{-1}
vacuum permeability	μ_0	$4\pi \times 10^{-7}$	H m^{-1}
vacuum permittivity	ϵ_0	$8.854\,187\,817 \cdots \times 10^{-12}$	F m^{-1}
vacuum impedance	Z_0	$376.730\,313\,461\ldots$	Ω
gravitational constant	G	$6.673(10) \times 10^{-11}$	$\text{m}^3\,\text{kg}^{-1}\,\text{s}^{-2}$
Planck constant	h	$6.626\,068\,76(52) \times 10^{-34}$	J s^{-1}
Planck mass	$m_\mathcal{P}$	$2.176\,7(16) \times 10^{-8}$	kg
Planck length	$l_\mathcal{P}$	$1.616\,0(12) \times 10^{-35}$	m
Planck time	$t_\mathcal{P}$	$5.390\,6(40) \times 10^{-44}$	s
Avogadro constant	N_A	$6.022\,141\,99(47) \times 10^{23}$	mol^{-1}
Bohr magneton	μ_B	$927.400\,899(37) \times 10^{-26}$	J T^{-1}
Bohr radius	a_0	$0.529\,177\,208\,3(19) \times 10^{-10}$	m
Boltzmann constant	k_B	$1.380\,650\,3(24) \times 10^{-23}$	J K^{-1}

continued on next page

Table 1.6: *continued*

QUANTITY	SYMBOL	VALUE	UNITS
elementary charge	e	$1.602\,176\,462(63) \times 10^{-19}$	C
Fine structure constant	α	$7.297\,352\,533(27) \times 10^{-3}$	
	α^{-1}	$137.035\,999\,76(50)$	
Gas constant	R	$8.314\,472(15)$	$\mathrm{J\,mol^{-1}\,K^{-1}}$
Nuclear magneton	μ_N	$5.050\,783\,17(20) \times 10^{-27}$	$\mathrm{J\,T^{-1}}$
Rydberg constant	R_∞	$10\,973\,731.568\,549(83)$	$\mathrm{m^{-1}}$
Stefan-Boltzmann constant	σ	$5.670\,400(40) \times 10^{-8}$	$\mathrm{W\,m^{-2}\,K^{-4}}$
Thomson cross section	σ_e	$0.665\,245\,854(15) \times 10^{-28}$	$\mathrm{m^2}$
Wien constant	b	$2.897\,768\,6(51) \times 10^{-3}$	m K
α particle:			
mass	m_α	$6.644\,655\,98(52) \times 10^{-27}$	kg
-electron mass ratio	m_α/m_e	$7.924\,299\,508(16) \times 10^3$	
-proton mass ratio	m_α/m_p	$3.972\,599\,684\,6(11)$	
deuteron:			
mass	m_d	$3.343\,583\,09(26) \times 10^{-27}$	kg
-electron mass ratio	m_d/m_e	$7.670\,482\,955\,0(78) \times 10^3$	
-proton mass ratio	m_d/m_p	$1.999\,007\,500\,83(41)$	
magnetic moment	μ_d	$4.330\,734\,57(18) \times 10^{-27}$	$\mathrm{J\,T^{-1}}$
electron:			
mass	m_e'	$9.109\,381\,88(72) \times 10^{-31}$	kg
-α particle mass ratio	m_e/m_α	$1.370\,933\,561\,1(29) \times 10^{-4}$	
-deuteron mass ratio	m_e/m_d	$2.724\,437\,117\,0(58) \times 10^{-4}$	
-proton mass ratio	m_e/m_p	$5.446\,170\,232(12) \times 10^{-4}$	

continued on next page

Table 1.6: *continued*

QUANTITY	SYMBOL	VALUE	UNITS
magnetic moment	μ_e	$-928.476\,362(37) \times 10^{-26}$	J T^{-1}
charge to mass ratio	$-e/m_e$	$-1.758\,820\,174(71) \times 10^{11}$	C kg^{-1}
classical radius	r_e	$2.817\,940\,285(31) \times 10^{-15}$	m
helion:			
mass	m_h	$5.006\,411\,74(39) \times 10^{-27}$	kg
-electron mass ratio	m_h/m_e	$5.495\,885\,238(12) \times 10^3$	
-proton mass ratio	m_h/m_p	$2.993\,152\,658\,50(93)$	
neutron:			
mass	m_n	$1.674\,927\,16(13) \times 10^{-27}$	kg
-electron mass ratio	m_n/m_e	$1.838\,683\,655\,0(40) \times 10^3$	
-proton mass ratio	m_n/m_p	$1.001\,378\,418\,87(58)$	
magnetic moment	μ_n	$-0.966\,236\,40(23) \times 10^{-26}$	J T^{-1}
proton:			
mass	m_p	$1.672\,621\,58(13) \times 10^{-27}$	kg
-electron mass ratio	m_p/m_e	$1.836\,152\,667\,5(39) \times 10^3$	
-neutron mass ratio	m_p/m_n	$0.998\,623\,478\,55(58)$	
magnetic moment	μ_p	$1.410\,606\,633(58) \times 10^{-26}$	$\text{J T}{-}1$

1.5 DIMENSIONAL ANALYSIS

Table 1.7: Dimensions of common variables

	SI		QUANTITY	CGS	
C	$\frac{q^2 t^2}{l^2 m}$	farad	**capacitance**	l	cm
q	q	coulomb	**charge**	$\frac{l^{3/2} m^{1/2}}{t}$	statcoulomb

continued on next page

Table 1.7: *continued*

	SI		QUANTITY	CGS	
ρ_c	$\frac{q}{l^3}$	coulomb m^{-3}	charge density	$\frac{m^{1/2}}{l^{3/2}t}$	statcoulomb cm^{-3}
S	$\frac{q^2 t}{l^2 m}$	siemens	conductance	$\frac{l}{t}$	cm s^{-1}
σ_c	$\frac{q^2 t}{l^3 m}$	siemens m^{-1}	conductivity	$\frac{1}{t}$	s^{-1}
I	$\frac{q}{t}$	ampere	current	$\frac{l^{3/2} m^{1/2}}{t^2}$	statampere
J	$\frac{q}{l^2 t}$	ampere m^{-2}	current density	$\frac{m^{1/2}}{l^{1/2} t^2}$	statampere cm^{-2}
D	$\frac{q}{l^2}$	coulomb m^{-2}	displacement	$\frac{m^{1/2}}{l^{1/2} t}$	statcoulomb cm^{-2}
η_v	$\frac{m}{lt}$	kg m^{-1}s^{-1}	dynamic viscosity	$\frac{m}{lt}$	poise
E	$\frac{ml}{qt^2}$	volt m^{-1}	electric field	$\frac{m^{1/2} l^{1/2}}{t}$	statvolt cm^{-1}
ϕ	$\frac{l^2 m}{qt^2}$	volt	electric potential	$\frac{l^{1/2} m^{1/2}}{t}$	statvolt
$\mathcal{E}$	$\frac{ml^2}{t^2}$	joule	energy	$\frac{ml^2}{t^2}$	erg
ϵ	$\frac{m}{lt^2}$	joule m^{-3}	energy density	$\frac{m}{lt^2}$	erg cm^{-3}
F	$\frac{lm}{t^2}$	newton	force	$\frac{lm}{t^2}$	dyne
ν	$\frac{1}{t}$	hertz	frequency	$\frac{1}{t}$	hertz
L	l	metre	length	l	cm
Φ	$\frac{l^2 m}{qt}$	weber	magnetic flux	$\frac{l^{3/2} m^{1/2}}{t}$	maxwell
B	$\frac{m}{qt}$	tesla	magnetic flux density	$\frac{m^{1/2}}{l^{1/2} t}$	gauss
H	$\frac{q}{lt}$	A m^{-1}	magnetic intensity	$\frac{m^{1/2}}{l^{1/2} t}$	oersted
μ	$\frac{l^2 q}{t}$	joule tesla^{-1}	magnetic moment	$\frac{l^{5/2} m^{1/2}}{t}$	oersted cm^3

continued on next page

Table 1.7: *continued*

	SI		QUANTITY	CGS	
m	m	kg	**mass**	m	gram
ρ	$\frac{m}{l^3}$	kg m^{-3}	**mass density**	$\frac{m}{l^3}$	gm cm^{-3}
P	$\frac{l^2 m}{t^3}$	watt	**power**	$\frac{l^2 m}{t^3}$	erg s^{-1}
p	$\frac{m}{lt^2}$	pascal	**pressure**	$\frac{m}{lt^2}$	dyne cm^{-2}
R	$\frac{l^2 m}{q^2 t}$	ohm	**resistance**	$\frac{t}{l}$	s cm^{-1}
η	$\frac{l^3 m}{q^2 t}$	ohm-m	**resistivity**	t	s
κ	$\frac{lm}{t^3}$	watt m^{-1}K^{-1}	**thermal conductivity**	$\frac{lm}{t^3}$	erg cm^{-1}s^{-1}K^{-1}
μ_0	$\frac{lm}{q^2}$	henry m^{-1}	**vacuum permeability**		
ϵ_0	$\frac{q^2 t^2}{l^3 m}$	farad m^{-1}	**vacuum permittivity**		
A	$\frac{lm}{qt}$	weber m^{-1}	**vector potential**	$\frac{l^{1/2} m^{1/2}}{t}$	gauss cm
u	$\frac{l}{t}$	m s^{-1}	**velocity**	$\frac{l}{t}$	cm s^{-1}

1.6 IONIZATION ENERGIES OF GAS-PHASE MOLECULES

The energies of first ionization E_i for certain gas-phase molecules are given here, selected from [57]

Table 1.8: Ionization energies of gas-phase molecules

SUBSTANCE	FORMULA	E_i/eV
Argon	Ar	15.75962
Carbon dioxide	CO_2	13.773
Carbon monoxide	CO	14.014
Chlorine	Cl	12.96764
Chlorine	Cl_2	11.480
Chlorosilane	ClH_3Si	11.4

continued on next page

Table 1.8: *continued*

SUBSTANCE	FORMULA	E_i/eV
Disodium	Na$_2$	4.894
Helium	He	24.58741
Hydrogen	H	13.59844
Hydrogen	H$_2$	15.42593
Hydrogen chloride	HCl	12.749
Krypton	Kr	13.999961
Mercury	Hg	10.43750
Neon	Ne	21.56454
Nitrogen	N	14.53414
Nitrogen	N$_2$	15.5808
Oxygen	O	13.61806
Oxygen	O$_2$	12.0697
Silane	SiH$_4$	11.00
Silicon	Si	8.15169
Sodium	Na	5.13908
Water	H$_2$O	12.6206
Xenon	Xe	12.12987

1.7 CHARACTERISTIC PARAMETERS FOR TYPICAL PLASMAS

Table 1.9: Operating parameters for rf parallel plate plasma etching, and high-density plasma reactor[9]

QUANTITY	RF	HIGH-DENSITY
pressure / Pa	$10^{-1} - 10^3$	$10^{-2} - 10$
pressure / torr	$0.001 - 10$	$10^{-4} - 10^{-1}$
power /W	$50 - 10^3$	$10^2 - 5 \times 10^3$
frequency /MHz	$0.1 - 100$	$0.1 - 20$ or 2.45GHz
gas flow rate / sccm [1]	$10 - 3 \times 10^3$	$10 - 200$
T_e/eV	$1 - 10$	$1 - 10$
plasma density /m^{-3}	$10^{14} - 10^{17}$	$10^{16} - 10^{19}$
fractional ionization	$10^{-7} - 10^{-4}$	$10^{-4} - 10^{-1}$
ion bombarding energy /eV	$50 - 10^3$	$10 - 500$

continued on next page

[1]standard cubic centimetres per second

Table 1.9: *continued*

QUANTITY	RF	HIGH-DENSITY
ion bombarding flux / mA cm^{-2}	$10^{-2} - 5$	$1 - 50$
magnetic field /T	0	$0 - 0.1$

Table 1.10: Ionospheric parameters [32]

IONOSPHERIC REGION	HEIGHT/km	n_e/m^{-3} (day)	n_e/m^{-3} (night)
D	50-90	10^9	10^8
E	90-140	10^{11}	$< 10^{10}$
F$_1$	140-200	3×10^{11}	10^{10}
F$_2$	200-400	10^{12}	10^{11}

Table 1.11: Solar plasma parameters [73, 99]

QUANTITY	REGION	TYPICAL VALUE
total number density /m^{-3}	photosphere	$10^{22} - 10^{23}$
electron number density /m^{-3}	photosphere	$10^{18} - 10^{20}$
total number density /m^{-3}	chromosphere	$10^{16} - 10^{22}$
electron number density /m^{-3}	chromosphere	$10^{16} - 10^{18}$
total number density /m^{-3}	corona	$10^8 - 10^{15}$
electron number density /m^{-3}	corona	$10^8 - 10^{15}$
temperature /K	photosphere	$4 \times 10^3 - 6 \times 10^3$
temperature /K	chromosphere	$\sim 3 \times 10^3 - 10^4$
temperature /K	corona	$> 10^6$
magnetic field strength /T	poles	$\sim 10^{-4}$
magnetic field strength /T	sunspot	~ 0.3
magnetic field strength /T	prominence	$10^{-3} - 10^{-2}$
magnetic field strength /T	chromospheric plage	$\sim 10^{-2}$

2

Basic Plasma Parameters

2.1 NOTATION

Symbol	Meaning	Ref
B	magnetic flux density	
c_a	Alfvén speed for the plasma	(2.24)
c_{as}	Alfvén speed for species s	(2.22)
c_{th}	gas sound speed	(2.25)
$c_{th,s}$	sound speed for gas species s	(2.25)
I	differential scattering cross-section	(2.29)
k_B	Boltzmann constant	
m_s	mass of particle of species s	
$\mathcal{M}$	Mach number	(2.42)
n_s	number density of particles of species s	
q_s	charge carried by particle of species s	
R_m	magnetic Reynolds number	(2.43)
s	label defining species: i (ion), e (electron), n (neutral)	
S	Lundquist number	(2.41)
T_s	temperature of gas of species s	
δ	plasma skin depth	(2.20)
ϵ_0	vacuum permittivity	
λ_D	Debye length	(2.17)
λ_{mfp}	mean free path of species n	(2.19)
μ_0	vacuum permeability	
μ_s	mobility of particle of species s	(2.34)
$\boldsymbol{\mu}_s$	mobility tensor for species s in a magnetised plasma	(2.36)
μ_{bs}	magnetic moment of a particle of species s	(2.33)
ν	non-specific collision frequency	
ν_{cs}	cyclotron frequency of species s (in Hz)	(2.9)
ν_{ps}	plasma frequency of species s (in Hz)	(2.3)
$\nu_{ss'}$	collision frequency for species s and s'	(2.12)
ρ_s	mass density of species s	
σ_{sc}	collision cross-section	(2.29)
τ_A	Alfvén transit time	(2.13)
τ_R	resistive diffusion time	(2.15)
ω	frequency of electromagnetic wave	
ω_{cs}	circular cyclotron frequency of species s	(2.7)
ω_p	circular plasma frequency	(2.6)
ω_{ps}	circular plasma frequency of species s	(2.1)

The following quantities are those natural time scales, length scales and other miscellaneous quantities which can be defined for a uniform plasma unconstrained by boundary conditions.

2.2 NATURAL TIMESCALES

2.2.1 Characteristic Frequencies

2.2.1.1 Plasma Frequency   The natural (circular) frequency of the collective oscillation of charged particles under a self-consistent electrostatic restoring force:

$$\omega_{ps} = \left(\frac{n_s q_s{}^2}{\epsilon_0 m_s}\right)^{1/2} \quad \text{(SI)} \tag{2.1}$$

$$= \left(\frac{4\pi n_s q_s{}^2}{m_s}\right)^{1/2} \quad \text{(cgs)} \tag{2.2}$$

Expressed as true frequencies, these formulae yield for electrons:

$$\nu_{pe} = \frac{\omega_{pe}}{2\pi} \tag{2.3}$$

$$\approx 9\sqrt{n_e}\,\text{Hz} \quad \text{(SI)} \tag{2.4}$$

$$\approx 9 \times 10^3 \sqrt{n_e}\text{s}^{-1} \quad \text{(cgs)} \tag{2.5}$$

Note that the plasma frequency of the whole plasma is given by

$$\omega_p = \left(\sum_s \omega_{ps}^2\right)^{1/2} \tag{2.6}$$

2.2.1.2 Cyclotron Frequency   The natural (circular) frequency of oscillation of charged particles in the presence of a magnetic field:

$$\omega_{cs} = \frac{q_s B}{m_s} \quad \text{(SI)} \tag{2.7}$$

$$= \frac{q_s B}{m_s c} \quad \text{(cgs)} \tag{2.8}$$

In terms of true frequencies, and for electrons,

$$\nu_{ce} = \frac{\omega_{ce}}{2\pi} \tag{2.9}$$

$$\approx \frac{28}{B}\text{GHz} \quad \text{(SI)} \tag{2.10}$$

$$\approx \frac{28 \times 10^{13}}{B}\text{s}^{-1} \quad \text{(cgs)} \tag{2.11}$$

Note that ω_{cs} takes the same sign as the charge on the particle.

2.2.1.3 Collision Frequency The collision frequency is the average rate at which inter-particle collisions take place, here assumed for simplicity to be between a mobile species (labelled with subscript s) and a stationary one. The general formula is given by

$$\nu_{cs} = n_n \sigma_{sc,s} \langle u_s \rangle \tag{2.12}$$

where n_n is the number density of stationary targets, $\sigma_{sc,s}$ is the collision cross-section, and $\langle u_s \rangle$ is the mean speed of the mobile species.

2.2.2 Characteristic Times

2.2.2.1 Alfvén Transit Time For an MHD plasma of typical dimension L, the Alfvén transit time τ_A is defined by

$$\tau_A = \frac{L}{c_a} \tag{2.13}$$

where c_a is the Alfvén speed, defined in (2.24).

2.2.2.2 Collision Time The general expression for the collision time τ_c defines it to be the reciprocal of the collision frequency:

$$\tau_c = \frac{1}{n_n \sigma_{sc} \langle u \rangle} \tag{2.14}$$

where $\langle u \rangle$ is the mean speed of the colliding particles. Specific definitions can be found in section 6.3.1.3.

2.2.2.3 Resistive Timescale For a resistive MHD plasma with characteristic dimension L, the resistive diffusion time τ_R is defined by

$$\tau_R = \frac{\mu_0}{\eta L^2} \quad \text{(SI)} \tag{2.15}$$

$$= \frac{4\pi L^2}{\eta c^2} \quad \text{(cgs)} \tag{2.16}$$

where η is the plasma resistivity.

2.3 NATURAL SCALELENGTHS

2.3.1 Debye Length

The exponential scale length for charge screening within an electron plasma with stationary ions is the Debye Length λ_D:

$$\lambda_D = \left(\frac{\epsilon_0 k_B T_e}{n_e e^2} \right)^{1/2} \qquad \text{(SI)} \qquad (2.17)$$

$$= \left(\frac{k_B T_e}{4\pi n_e e^2} \right)^{1/2} \qquad \text{(cgs)} \qquad (2.18)$$

This expression can be generalised to define a Debye length for each species s, but is usually reserved for electrons.

2.3.2 Mean Free Path

The mean free path is the average distance a particle moves before successive collisions (or interactions); it is also therefore the exponential scale factor for the spatial decay of particle flux as a result of collisions. It is defined by λ_{mfp}:

$$\lambda_{\mathrm{mfp}} = (n_n \sigma_{sc})^{-1} \qquad (2.19)$$

where n_n is the neutral number density, and σ_{sc} is the collision cross-section (see (2.29)).

2.3.3 Plasma Skin Depth

The spatial decay constant for electromagnetic radiation of frequency ω incident on the boundary of a uniform density plasma of dielectric constant ϵ:

$$\delta = \frac{c}{\omega_{pe}} [Im(\epsilon^{1/2})] \qquad (2.20)$$

See (2.37), (7.135) for examples of plasma dielectric constants. Note that magnetized plasmas are anisotropic, and have dielectric tensors; see (7.20, 7.152) for details.

2.3.4 Larmor Radius

The radius of the circular orbit of a charged particle in the plane perpendicular to a uniform magnetic field r_{Ls} :

$$r_{Ls} = \frac{v_{\perp s}}{\omega_{cs}} \qquad (2.21)$$

where $v_{\perp s}$ is the speed in the plane of a particle of species s.

2.4   NATURAL SPEEDS

2.4.1   Alfvén Speed

The speed typically at which magnetic disturbances are propagated by particles of species s is given by

$$c_{as} = \left(\frac{B^2}{\mu_0 \rho_s}\right)^{1/2} \quad \text{(SI)} \tag{2.22}$$

$$= \left(\frac{B^2}{4\pi \rho_s}\right)^{1/2} \quad \text{(cgs)} \tag{2.23}$$

The Alfvén speed for the whole plasma is given by

$$c_a = \left(\sum_s c_{as}^2\right)^{1/2} \tag{2.24}$$

2.4.2   Sound Speed

The speed of sound $c_{th,s}$ in a gas of particles of species s is defined by

$$c_{th,s}^2 = \left(\frac{\mathrm{d}p_s}{\mathrm{d}\rho_s}\right)_{S_0} \tag{2.25}$$

$$= \gamma \frac{p_{s0}}{\rho_{s0}} \tag{2.26}$$

where p_s is the gas pressure, ρ_s the mass density, and subscript S_0 denotes that the derivative is taken at constant entropy; γ is the polytropic index, defined by the equation of state

$$p_s \rho_s^{-\gamma} = \quad \text{constant} \tag{2.27}$$

An isothermal gas has $\gamma = 1$; an adiabatic one has $\gamma = 5/3$. For an ideal gas, $p = nk_BT$ where n is the particle number density, and T is the temperature. Consequently an alternative form of the sound speed is

$$c_{th,s} = \left(\frac{k_B T_s}{m_s}\right)^{1/2} \tag{2.28}$$

2.5   MISCELLANEOUS PARAMETERS

2.5.1   Collision Cross-Section

There are several quantities which can be defined as collision, or scattering, cross-sections. The total scattering cross-section σ_{sc} is defined as

$$\sigma_{sc} = 2\pi \int_0^\pi I(v, \theta) d\theta \qquad (2.29)$$

where $I(v, \theta)$ is the differential scattering cross-section, v is the relative speed of the scattered particle compared with the target, and θ is the angle through which the particle is scattered by the collision. Where the particles are hard, elastic spheres of radii a_1 and a_2, then

$$\sigma_{sc} = \pi(a_1 + a_2)^2 \qquad (2.30)$$

2.5.2   Differential Scattering Cross-Section

The differential scattering cross-section is defined to be the energy radiated per unit time, per unit solid angle, divided by the incident energy flux.

For Coulomb collisions between charged particles, an elastic scattering process, the differential cross-section for scattering into unit solid angle is given by the Rutherford formula

$$I(v_0, \Theta) = \frac{Z_1{}^2 Z_2{}^2 e^4}{(8\pi\epsilon_0)^2 m_R{}^2 v_0^4 \sin^4(\Theta/2)} \quad \text{(SI)} \qquad (2.31)$$

$$= \frac{Z_1^2 Z_2^2 e^4}{4 m_R^2 v_0^4 \sin^4(\Theta/2)} \quad \text{(cgs)} \qquad (2.32)$$

where $Z_i e$ is the charge on particle i, $m_R = m_1 m_2/(m_1 + m_2)$ is the reduced mass, v_0 is the relative speed, and Θ is the collision angle in the centre of mass frame.

2.5.3   Magnetic Moment

The magnetic moment of a charged particle performing Larmor orbits in a magnetised plasma:

$$\mu_s = \frac{m_s v_{\perp s}{}^2}{2B} \qquad (2.33)$$

2.5.4   Mobility

The mobility of a plasma particle of species s in a collisional plasma is defined in simple terms as the magnitude of the mean plasma particle flow produced per unit strength of applied steady electric field:

$$\mu_s = \frac{|q_s|}{m_s \nu}.$$

(2.34)

Where an external AC electric field of frequency ω is applied, the particle mobility can be re-defined as

$$\mu_s = \frac{|q_s|}{m_s(\nu + i\omega)}$$

(2.35)

If the plasma also has an applied magnetic field, then the particle mobility becomes a tensor:

$$\boldsymbol{\mu_s} = \frac{q_s}{m} \frac{1}{(\nu + i\omega)^2 + \omega_{cs}^2} \times$$

$$\begin{bmatrix} \nu + i\omega & \omega_{cs} & 0 \\ -\omega_{cs} & \nu + i\omega & 0 \\ 0 & 0 & \dfrac{(\nu + i\omega)^2 + \omega_{cs}^2}{\nu + i\omega} \end{bmatrix}$$

(2.36)

where the magnetic field is taken to lie along the z-axis.

2.6 NON-DIMENSIONAL PARAMETERS

2.6.1 Dielectric Constant

A cold, unmagnetised plasma has a frequency dependent relative dielectric constant given by

$$\epsilon = \left(1 - \frac{\omega_p^2}{\omega(\omega + i\nu_{en})}\right)^{1/2}$$

(2.37)

where ν_{en} is the electron-neutral collision frequency.

A magnetised plasma has a dielectric tensor, reflecting the intrinsic anisotropy caused by the magnetic field. The detailed description of the dielectric tensor depends crucially on the modelling assumptions: see Section 7.2.3 for the cold plasma model, and Section 7.4.4 for a kinetic treatment.

2.6.2 Hartmann Number

The Hartmann number, $\mathcal{H}_a$, is the ration of the magnetic force to the dissipative force, and is defined by

$$\mathcal{H}_a = \frac{BL}{(\eta\eta_v)^{1/2}} \qquad \text{(SI)} \tag{2.38}$$

$$\mathcal{H}_a = \frac{BL}{c\,(\eta\eta_v)^{1/2}} \qquad \text{(cgs)} \tag{2.39}$$

where B is the magnetic induction, η_v is the fluid viscosity, η is the resistivity, and L is a typical scale-length of the system.

2.6.3 Knudsen Number

The Knudsen number $\mathcal{K}_n$ is defined to be the ratio of the mean-free-path of the gas and the characteristic scalelength L of the gas volume:

$$\mathcal{K}_n = \frac{\lambda_{\mathrm{mfp}}}{L} \tag{2.40}$$

$\mathcal{K}_n$ is very small for collisionally dominated confined gases, but can rise to near unity for some low pressure discharges. If $\mathcal{K}_n > 1$, the flow is termed (free) molecular flow; $\mathcal{K}_n < 0.01$ describes viscous flow; and $0.01 < \mathcal{K}_n \leq 1$ characterises transitional flows.

2.6.4 Lundquist Number

The Lundquist number S for a resistive MHD plasma is the ratio of the timescales for diffusive processes to that for dynamical processes, and is defined by

$$S = \frac{\tau_R}{\tau_A} \tag{2.41}$$

where τ_R and τ_A are defined by (2.15) and (2.13) respectively.

2.6.5 Mach Number

The ratio of fluid speed u to the fluid sound speed c_{th} is termed the Mach number:

$$\mathcal{M} = u/c_{th} \tag{2.42}$$

2.6.6 Magnetic Reynolds Number

The Reynolds number in fluid mechanics is the ratio of the inertial to viscous forces. In a magnetised plasma of resistivity η, a magnetic Reynolds number can be defined in an analogous way:

$$R_m = \frac{\mu_0 u L}{\eta} \qquad \text{(SI)} \qquad\qquad (2.43)$$

$$= \frac{u L}{\eta} \qquad \text{(cgs)} \qquad\qquad (2.44)$$

where u and L are a characteristic speed and length scale, respectively.

2.6.7   Plasma Beta

For an MHD plasma, the plasma beta (β) is defined as the ratio of thermo-dynamic pressure to magnetic pressure:

$$\beta = \frac{p}{B^2/(2\mu_0)} \qquad \text{(SI)} \qquad\qquad (2.45)$$

$$= \frac{p}{B^2/(8\pi)} \qquad \text{(cgs)} \qquad\qquad (2.46)$$

3

Discharge Plasmas and Elementary Processes

3.1 NOTATION

SYMBOL	MEANING	REF
B	magnetic flux density	
d	electrode separation	
d_s	planar sheath extent	
D_a	ambipolar diffusion coefficient	(3.42)
D_s	diffusion coefficient for species s	(3.32)
E	electric field	
i_0	primary electron current at cathode	(3.56)
i_a	electron current at anode	(3.56)
J_i	ion current density	(3.9)
m_s	mass of particle of species s	
n_s	number density of particles of species s	
N_e	total number of electrons	(3.55)
N_{e0}	total number of electrons emitted at cathode	(3.55)
p	neutral gas pressure	(3.60)
q_s	charge carried by particle of species s	
s	label defining species: i (ion), e (electron), n (neutral)	
T_s	temperature of gas of species s	
u_0	ion speed at the plasma-sheath edge	(3.4)
u_i	ion speed in the sheath	(3.3)
V	voltage	
V_b	breakdown voltage	(3.73)
$V_{b,min}$	minimum breakdown voltage	(3.77)
α_{T}	first Townsend ionization coefficient	(3.56)
γ_{T}	second Townsend ionization coefficient	(3.67)
$\boldsymbol{\Gamma}_s$	flux of particles of species s	(3.28)
δ	plasma skin depth	(2.20)
ϵ_0	vacuum permittivity	
λ_D	Debye length	(2.17)
λ_{mfp}	mean free path	(2.19)
μ_s	mobility of particle of species s	(2.34)
$\boldsymbol{\mu}_s$	mobility tensor for species s in a magnetised plasma	(2.36)
ν	non-specific collision frequency	
ν_{cs}	collision frequency of species s (in Hz)	
σ_{sc}	collision cross-section	(2.29)
ω	frequency of electromagnetic wave	
ω_{cs}	circular cyclotron frequency of species s	(2.7)
ω_p	circular plasma frequency	(2.6)
ω_{ps}	circular plasma frequency of species s	(2.1)

3.2 PLASMA SHEATH

In very general terms, a plasma bounded by an absorbing wall loses mobile electrons to the wall, and shields itself from the resulting electric field by the creation of a positive space charge region, termed the sheath.

3.2.1 Planar Sheath Equation

The standard model of a free-fall stationary planar sheath is presented, observing the following assumptions:

- ions are cold

- electrons obey Boltzmann statistics in a 1-dimensional model

- the sheath extent is small enough for the sheath to be collisionless, though the plasma need not be

In this model, the equilibrium electron and ion number densities, and the ion speed, as a function of sheath distance x, are given by:

$$n_e(x) = n_0 \exp e\phi/(k_B T_e) \tag{3.1}$$

$$n_i(x) = n_0 \left(1 - \frac{2e\phi}{m_i u_0^2} \right)^{-1/2} \tag{3.2}$$

$$u_i(x) = \left(u_0^2 - \frac{2e\phi}{m_i} \right)^{1/2} \tag{3.3}$$

$$u_0 = u_i(x = 0) \tag{3.4}$$

The non-linear equation for the structure of the electric potential across the sheath is

$$\frac{d^2\phi}{dx^2} = \frac{en_0}{\epsilon_0} \left[\exp\left(\frac{e\phi}{k_B T_e} \right) - \left(1 - \frac{2e\phi}{m_i u_0^2} \right)^{-1/2} \right] \quad \text{(SI)} \tag{3.5}$$

$$= 4\pi e n_0 \left[\exp\left(\frac{e\phi}{k_B T_e} \right) - \left(1 - \frac{2e\phi}{m_i u_0^2} \right)^{-1/2} \right] \quad \text{(cgs)} \tag{3.6}$$

where we take $x = 0$ to be the sheath-plasma interface, at which ϕ and $d\phi/dx$ are assumed to be zero.

3.2.1.1 Bohm Sheath Criterion   Note that (3.5) has monotonic solutions for the potential (avoiding trapped ions) only if

$$u_0 > \left(\frac{k_B T_e}{m_i} \right)^{1/2} \tag{3.7}$$

which is equivalent to demanding that the ion number density falls more slowly than the electron number density across the sheath, allowing the positive space-charge shield to develop. This concept can be generalised as [77]

$$
\left[\frac{\mathrm{d}}{\mathrm{d}\chi}(n_i - n_e) \geq 0\right]_{\chi=0}
\tag{3.8}
$$

with $\chi = -e\phi/(k_B T_e)$. Note that (3.7) and (3.8) demand that the cold ions are accelerated before entering the sheath region; this necessitates a so-called presheath region in which the requisite acceleration mechanism is present.

Note that the Bohm criterion applies strictly only when the mean free path for particles in the sheath is much greater than the sheath extent, so that the sheath is collisionless (but the plasma needn't necessarily be). It does not have to be satisfied if the sheath is collisional, that is, if the local Debye length is greater than the ion mean free path [77].

3.2.2   Child-Langmuir Law

The Child-Langmuir law gives the space-charge limited ion current density in a planar sheath of width d:

$$
J_i = \frac{4}{9}\epsilon_0 \left(\frac{2e}{m_e}\right)^{1/2} \frac{\phi_0^{3/2}}{d_s^2} \quad \text{(SI)}
\tag{3.9}
$$

$$
= \frac{1}{9\pi} \left(\frac{2e}{m_e}\right)^{1/2} \frac{\phi_0^{3/2}}{d_s^2} \quad \text{(cgs)}
\tag{3.10}
$$

The following restrictions on the validity of this result apply:

- sheath is collisionless

- the electron number density is ignored when solving (3.5)

- the ion current J_i is constant across the sheath

- $e\phi/(k_B T_e) \gg 1$

- (3.9) strictly only applies close to the wall

Assuming the Child-Langmuir law (3.9), we have:

$$d_s = \frac{2^{1/2}}{3} \lambda_D \left(\frac{2\phi_0}{T_e} \right) \tag{3.11}$$

$$\phi = -\phi_0 (x/d_s)^{4/3} \tag{3.12}$$

$$E = \frac{4\phi_0}{3d_s} \left(\frac{x}{d_s} \right)^{1/3} \tag{3.13}$$

$$n_i = \frac{4\epsilon_0}{9e} \frac{\phi_0}{d_s^2} \left(\frac{x}{d_s} \right)^{-2/3} \quad \text{(SI)} \tag{3.14}$$

$$n_i = \frac{16\pi}{9e} \frac{\phi_0}{d_s^2} \left(\frac{x}{d_s} \right)^{-2/3} \quad \text{(cgs)} \tag{3.15}$$

3.2.3 Collisional Sheaths

If the collisional scale-length for ions is less than the sheath extent d_s then the form of the sheath potential is modified [58]:

$$\phi = -\frac{3}{5} \left(\frac{3}{2\epsilon_0} \right)^{2/3} \frac{(en_0 u_0)^{2/3}}{[2e\lambda_{\mathrm{mfp},i}/(\pi m_i)]^{1/3}} x^{5/3} \quad \text{(SI)} \tag{3.16}$$

$$= -\frac{3}{5} (6\pi)^{2/3} \frac{(en_0 u_0)^{2/3}}{[2e\lambda_{\mathrm{mfp},i}/(\pi m_i)]^{1/3}} x^{5/3} \quad \text{(cgs)} \tag{3.17}$$

This can be rearranged to yield the collisional form of the Child-Langmuir law:

$$J_i = \frac{2}{3} \left(\frac{5}{3} \right)^{3/2} \epsilon_0 \left(\frac{2e\lambda_{\mathrm{mfp},i}}{\pi m_i} \right)^{1/2} |\phi_0|^{3/2} d_s^{-5/2} \quad \text{(SI)} \tag{3.18}$$

$$= \frac{2}{3} \left(\frac{5}{3} \right)^{3/2} (4\pi)^{-1} \left(\frac{2e\lambda_{\mathrm{mfp},i}}{\pi m_i} \right)^{1/2} |\phi_0|^{3/2} d_s^{-5/2} \quad \text{(cgs)} \tag{3.19}$$

where $\lambda_{\mathrm{mfp},i}$ is assumed independent of ion speed, and ϕ_0 is the potential at the electrode.

Note that (3.16), (3.18) depend implicitly on defining a plasma-sheath edge. A piecewise continuous modelling approach to accommodating a Bohm criterion with a collisional sheath is used in [39], allowing a plasma-sheath edge to be defined. However a matched-asymptotic expansion approach to modelling the plasma and sheath [35] suggests that there is a transition layer joining the plasma to the free-fall sheath: the transition region scales as $\lambda_{D0}^{8/9}$, where λ_{D0} is the debye length evaluated at the central electron density in the discharge, and the potential across the transition region varies as $\lambda_{D0}^{4/9}$. The existence of such a transition region suggests that the identification of a sharp plasma-sheath boundary may be problematical.

3.3   DOUBLE-LAYER

An analogous phenomenon to the wall sheath is the double-layer (DL), which
is an isolated electrostatic structure in a current-carrying plasma, and which
though overall charge neutral, sustains a significant potential difference. The
DL acts as a potential barrier to certain particles, for which the DL potential
is too great for them to overcome, and they are reflected. Conversely, particles
which do manage to cross are accelerated; those accelerated to higher energies
can emerge from the DL as a particle beam.

The DL structure is determined self-consistently from the disposition of
charged particles in an electric field, usually requiring populations of reflected
and accelerated particles. The current is carried by the free (non-reflected
traversing) particles, mainly by electrons in the non-relativistic description,
but evenly by ions and electrons for the relativistic DL.

The mathematical description [74] of the DL depends upon Poisson's equation:

$$-\epsilon_0 \phi'' = \frac{i_i}{[(2e\phi_{\mathrm{DL}} - \phi)/m_i]^{1/2}} - \frac{i_e}{(2e\phi/m_e)^{1/2}} \quad \text{(SI)} \qquad (3.20)$$

$$-\frac{\phi''}{4\pi} = \frac{i_i}{[(2e\phi_{\mathrm{DL}} - \phi)/m_i]^{1/2}} - \frac{i_e}{(2e\phi/m_e)^{1/2}} \quad \text{(cgs)} \qquad (3.21)$$

subject to $\phi(0) = \phi_{\mathrm{DL}}$, $\phi(d) = 0$, where ϕ_{DL} is the potential drop in the double
layer, i_s is the current of species s, and d is the DL extent. Note that only
one species of ion is considered here. The solution to (3.20) is a variation of
the Child-Langmuir law:

$$(i_e + i_i)d^2 = \frac{4}{9}\epsilon_0 C_0 \left(\frac{2e}{m_i}\right)^{1/2} \left[1 + (m_e/m_i)^{1/2}\right] \phi_{\mathrm{DL}}^{3/2} \quad \text{(SI)} \qquad (3.22)$$

$$= \frac{1}{9\pi} C_0 \left(\frac{2e}{m_i}\right)^{1/2} \left[1 + (m_e/m_i)^{1/2}\right] \phi_{\mathrm{DL}}^{3/2} \quad \text{(SI)} \qquad (3.23)$$

$$C_0 \approx 1.867 \qquad (3.24)$$

for which

$$\frac{i_e}{i_i} = \left(\frac{m_i}{m_e}\right)^{1/2} \qquad (3.25)$$

The above results are for a non-relativisitic DL. The relativistic equivalents are

$$
\begin{cases}
(i_e + i_i)d^2 \approx \tfrac{1}{2}\pi^2 \epsilon_0 c\phi_{\mathrm{DL}} \left[1 + \left(\dfrac{e\phi_{\mathrm{DL}}}{2m_i c^2}\right)^{1/2}\right] & m_e c^2 \ll e\phi_{\mathrm{DL}} \ll m_i c^2 \quad (\mathrm{SI}) \\[4mm]
(i_e + i_i)d^2 \approx \tfrac{\pi}{8} c\phi_{\mathrm{DL}} \left[1 + \left(\dfrac{e\phi_{\mathrm{DL}}}{2m_i c^2}\right)^{1/2}\right] & m_e c^2 \ll e\phi_{\mathrm{DL}} \ll m_i c^2 \quad (\mathrm{cgs}) \\[4mm]
(i_e + i_i)d^2 \approx \dfrac{4\epsilon_0 c e}{m_i c^2}\phi_{\mathrm{DL}}^2 & e\phi_{\mathrm{DL}} > m_i c^2 \quad (\mathrm{SI}) \\[4mm]
(i_e + i_i)d^2 \approx \dfrac{c e}{\pi m_i c^2}\phi_{\mathrm{DL}}^2 & e\phi_{\mathrm{DL}} > m_i c^2 \quad (\mathrm{cgs})
\end{cases}
$$

$$(3.26)$$

The extension to the Langmuir current condition (3.25) is

$$
\frac{i_e}{i_i} = \left(\frac{2m_i c^2 + e\phi_{\mathrm{DL}}}{2m_e c^2 + e\phi_{\mathrm{DL}}}\right)^{1/2} \tag{3.27}
$$

These static analytical DL solutions are only special simple cases; the general solution has to be numerical. Consult [74] for a comprehensive review of more realistic approaches.

3.4 DIFFUSION PARAMETERS

3.4.1 Free Diffusion

For a neutral gas, the flux $\mathbf{\Gamma}_s$ of particles of species s is given by [18, 64]

$$
\mathbf{\Gamma}_s = -\nabla \left(n_s \left\langle \tfrac{1}{3} v_s^2 \nu_{cs} \right\rangle\right) \tag{3.28}
$$

$$
= -\tfrac{1}{3} \langle v_s^2 \rangle \lambda_{\mathrm{mfp},s} \nabla n_s \tag{3.29}
$$

where $\langle \cdots \rangle$ denotes the average value, v_s is the particle speed, ν_{cs} is the particle collision frequency, $\lambda_{\mathrm{mfp},s}$ is the mean free path, and n_s the number density. The diffusion equation can be written in the form

$$
\frac{\partial n_s}{\partial t} + \nabla \cdot \mathbf{\Gamma}_s = 0 \tag{3.30}
$$

assuming no sources or sinks of particles. For parameters that are constant in space, (3.30) can be written as [18]

$$
\frac{\partial n_s}{\partial t} = D_s \nabla^2 n_s \tag{3.31}
$$

$$
D_s = \frac{k_B T}{m_s \nu_{cs}} \tag{3.32}
$$

where m_s is the mass of a particle of species s. Written in this form, (3.31) and (3.32) are referred to as Fick's Law of diffusion.

3.4.2   Mobility

The mobility μ_s of a charged particle is defined in terms of the drift speed produced by an applied electric field. Hence [18]

$$\mu_s = \frac{|q_s|}{m_s \nu_{cs}} \qquad \text{dc electric field} \qquad (3.33)$$

$$= \frac{|q_s|}{m_s(\nu_{cs} + i\omega)} \qquad \text{ac electric field} \qquad (3.34)$$

where the particle has charge q_s and mass m_s, and where ν_{cs} is the collision frequency. In the ac-case, the applied electric field has frequency ω. Note that since

$$\nu_{cs} = \langle u \rangle / \lambda_{\mathrm{mfp}} \qquad \text{from (2.12)} \qquad (3.35)$$

$$= \langle u \rangle n_n \sigma_{sc} \qquad \text{from (2.19)} \qquad (3.36)$$

then

$$\mu_s = \frac{|q|}{m_s n_n \sigma_{sc} \langle u \rangle} \qquad (3.37)$$

$$\propto \frac{1}{p} \qquad (3.38)$$

where $p = n_n k_B T_g$ is the ideal gas law for the neutral gas, number density n_n and temperature T_g. Consequently,

$$v_{ds} \propto \frac{E}{p} \qquad (3.39)$$

where v_{ds} is the drift speed of particles of species s.
  See also Section 2.5.4.

3.4.3   Ambipolar Diffusion

Where the number density of charged particles is sufficiently large ($n_e \approx n_i \approx 10^{14}\,\mathrm{m^{-3}} \approx 10^8\,\mathrm{cm^{-3}}$) that their mutual coulomb field affects their transport, the free-diffusion assumptions of Fick's Law ((3.31) and (3.30)) must be modified. In such circumstances the particle flux $\boldsymbol{\Gamma}_s$ may be written in terms of the diffusion and mobility parameters:

$$\boldsymbol{\Gamma}_s = \pm \mu_s n_s \boldsymbol{E} - D_s \nabla n_s \qquad (3.40)$$

where μ_s is the particle mobility ((3.33) and (3.34)), n_s the number density, $\boldsymbol{E}$ the electric field, and D_s the diffusion coefficient (3.32).

The common flux $\boldsymbol{\Gamma}$ of ions and electrons in the presence of an electric field can be written in a form analogous to (3.31), defining the ambipolar diffusion coefficient D_a:

$$\boldsymbol{\Gamma} = D_a \nabla n \tag{3.41}$$

$$D_a = -\frac{\mu_i D_e + \mu_e D_i}{\mu_i + \mu_e} \tag{3.42}$$

$$\approx D_i \left(1 + \frac{T_e}{T_g}\right) \tag{3.43}$$

where T_e and T_g are the electron and neutral gas temperatures, respectively.

The electric field $\boldsymbol{E}_s$ of the space-charge which results from the faster ambipolar diffusion of electrons can be quantified as

$$\boldsymbol{E}_s = -\frac{D_e - D_i}{\mu_e + \mu_i}\frac{\nabla n_s}{n_s} \tag{3.44}$$

For more than one species of positive ion, the ambipolar diffusion coefficients of the ions are unchanged, but the electron diffusion is altered [18]:

$$D_{aj} \approx D_{ij}\left(1 + \frac{T_e}{T_g}\right) \qquad j = 1, 2, \ldots, N_i \tag{3.45}$$

$$D_{ae} \approx \frac{1}{n_e}\sum_{j=1}^{N_i} n_{ij} D_{aj} \tag{3.46}$$

where N_i is the total number of ion species present.

For a gas containing negative ions,

$$D_{a+} \approx D_i\left(1 + \frac{T_e}{T_g}\right) \tag{3.47}$$

$$D_{a-} \approx 2\left(1 + \frac{n_{i-}}{n_e}\right)\frac{D_i D_{i-}}{D_e} - D_{i-}\left(\frac{T_e}{T_g} - 1\right) \tag{3.48}$$

$$D_{ae} \approx \left(1 + \frac{n_{i-}}{n_e}\right)D_+\left(1 + \frac{T_e}{T_g}\right) + \frac{n_{i-}}{n_e}D_{i-}\left(\frac{T_e}{T_g} - 1\right) \tag{3.49}$$

where $D_{a\pm}$ is the ambipolar diffusion coefficient for the positive and negative ions, respectively, and D_{ae} is that for electrons, and subscript $i-$ refers to the negative ions.

3.4.3.1 Restrictions

- $n_i = n_e = n$ is assumed

- (3.42) assumes a steady state, that is, no time evolution

- the mobility and diffusion coefficients are assumed to be constant in space, and independent of energy

- no particle sources or sinks are present

- electrons and ions move at a common speed: $v_i = v_e = v$

3.4.4   Ambipolar Diffusion in a Magnetic Field

Here the particle flux $\boldsymbol{\Gamma}_s$ is given by

$$\boldsymbol{\Gamma}_s = \pm n_s \mu_s \boldsymbol{E} \pm \mu_s (\boldsymbol{\Gamma}_s \times \boldsymbol{B}) - D_s \nabla n_s \qquad (3.50)$$

Particle motion can be split into two cases: parallel to $\boldsymbol{B}$, and perpendicular to $\boldsymbol{B}$. For the parallel case, the mobility and diffusion are unaffected by the magnetic field:

$$\boldsymbol{\Gamma}_{\|s} = \pm \mu_s n_s E_{\|} - D_s \nabla_{\|} n_s \qquad (3.51)$$

where μ_s and D_s are as before, and $\nabla_{\|}$ denotes the derivative along the direction of $\boldsymbol{B}$.

The motion perpendicular to $\boldsymbol{B}$ is affected, with the perpendicular flux given by

$$\boldsymbol{\Gamma}_{\perp s} = \pm \mu_{\perp s} n_s \boldsymbol{E}_{\perp} - D_{\perp s} \nabla_{\perp} n_s \qquad (3.52)$$

where

$$\mu_{\perp s} = \frac{\mu_s}{1 + \omega_{cs}^2 / \nu_{cs}^2} \qquad (3.53)$$

$$D_{\perp s} = \frac{D_s}{1 + \omega_{cs}^2 / \nu_{cs}^2} \qquad (3.54)$$

3.4.4.1   Restrictions   The same restrictions apply here as in Section 3.4.3.1, with the additional constraint that the magnetic field is assumed uniform in space.

3.5   IONIZATION

3.5.1   Townsend Breakdown

An electric field E applied to a gas with some seed ionization already present (from cosmic rays, for example) will yield a current which increases with electric field, provided the electric field imparts to electrons energies higher than the ionization potential of the gas.

3.5.1.1 Townsend's First Ionization Coefficient Townsend's formula for this process is [18, 45, 59]

$$N_e = N_{e0}e^{\alpha_{\mathrm{T}}x} \qquad (3.55)$$

where N_e is the number of electrons at a distance x from the cathode, N_{e0} is the number of electrons emitted at the cathode, and α_{T} is the number of ionizing collisions made per unit length, (equivalently, the number of ion-pairs produced per electron per unit drift length) known as *Townsend's first ionization coefficient*. Note that α_{T} is a function of the gas composition.

If the electrode separation is d, then the current at the anode due solely to electron creation via the first Townsend ionization coefficient, neglecting diffusion losses, is

$$i_a = i_0 e^{\alpha_{\mathrm{T}}d} \qquad (3.56)$$

where i_0 is the primary electron current at the cathode, and where $i_a - i_0$ is the positive ion current at the cathode. The quantity i_a/i_0 is known as the *multiplication factor*.

An alternative approach is to consider the number of ionizing collisions per unit voltage difference [18]:

$$N_e = N_{e0} \exp[\bar{\eta}(V - V_0)] \qquad (3.57)$$

$$\bar{\eta} = \frac{1}{V - V_0} \int_0^V \eta(V')\mathrm{d}V' \qquad (3.58)$$

$$\eta = \frac{\alpha_{\mathrm{T}}}{E} \qquad (3.59)$$

where V is the voltage, and V_0 is the threshold voltage for the effect to be seen.

Since the mean distance between electron-neutral collisions is λ_{mfp}, each electron drifting in an electric field must gain energy $eE\lambda_{\mathrm{mfp}}$ after each collision. Since α_{T} is the number of ionizing collisions per unit length, then it is reasonable to assume that α_{T} must be a function of the neutral gas pressure (number of encounters per unit length) and energy gain per collision. Hence

$$\frac{\alpha_{\mathrm{T}}}{p} = \mathcal{F}\left(\frac{E}{p}\right) \qquad (3.60)$$

$$= C_1 \exp\left(-C_2 \frac{p}{E}\right) \qquad (3.61)$$

for some function $\mathcal{F}$; the form (3.61) is due to Townsend, with C_1, C_2 constants which depend on the neutral gas. Values of C_1 and C_2 for various gases are given in Table 3.2 [79], with curves of the first ionization coefficient based on (3.61) shown for some of these gases in Figure 3.1. The Townsend formula

Table 3.2 The values of C_1 and C_2 for the analytical model of the first Townsend ionization coefficient, given in (3.61). Note that the first two data columns are derived from the last two, using the unit conversions $1\,\mathrm{cm} = 0.01\mathrm{m}$, and $1\,\mathrm{torr} = 1\,\mathrm{mm}\text{-}\mathrm{Hg} = 133.3224\,\mathrm{Pa}$ [96]. The data for gases marked with * may be too high by up to a factor 2 [79]. These data are valid generally in the range $C_2/2 \le E/p \le 3C_2$, but see also the noble gas model ((3.62) and Table 3.3).

Gas	C_1 $(\mathrm{m^{-1}\,Pa^{-1}})$	C_2 $(\mathrm{V\ m^{-1}Pa^{-1}})$	C_1 $(\mathrm{cm^{-1}\,torr^{-1}})$	C_2 $(\mathrm{V\ cm^{-1}torr^{-1}})$
H_2	7.95	263	10.6	350
N_2	9.0	256	10.6	342
CO^*_2	15	350	20	466
Air	9.15	274	12.2	365
H_2O	9.68	217	12.9	289
HCl*	18.8	285	25	380
Hg	15	278	20	370
He	1.37	37.5	1.82	50
Ne	3	75	4	100
Ar	9.0	150	12	200
Kr	10.9	165	14.5	220
Xe	16.7	233	22.2	310

(3.61) can be extended to better account for the noble gases [94]:

$$\frac{\alpha_\mathrm{T}}{p} = D_1 \exp\left[-D_2\left(\frac{p}{E}\right)^{\frac{1}{2}}\right] \qquad (3.62)$$

where the square-root dependence is empirical. The appropriate coefficients for this model are given in Table 3.3.

The standard Townsend formula (3.61) can be extended to incorporate the effect of a magnetic field [18]:

$$\frac{\alpha_\mathrm{T}}{p} = C_1\left(\frac{\nu_{cs}^2 + \omega_{cs}^2}{\nu_{cs}^2}\right) \exp\left[-C_2\frac{p}{E}\left(\frac{\nu_{cs}^2 + \omega_{cs}^2}{\nu_{cs}^2}\right)^{\frac{1}{2}}\right] \qquad (3.63)$$

3.5.1.2 Stoletow Point It is known experimentally that there is a pressure for which the multiplication at fixed voltage is a maximum, that is,

$$\frac{\partial \alpha_\mathrm{T}}{\partial p} = 0 \qquad (3.64)$$

Table 3.3 The values of D_1 and D_2 for the analytical model of the first Townsend ionization coefficient for noble gases, given in (3.62)[94]. Note that the first two data columns are derived from the last two, using the unit conversions $1\,\mathrm{cm} = 0.01\mathrm{m}$, and $1\,\mathrm{torr} = 1\,\mathrm{mm\text{-}Hg} = 133.3224\,\mathrm{Pa}$ [96]. The uncertainty in these data is approximately 7%.

Gas	D_1 $(\mathrm{m^{-1}\,Pa^{-1}})$	D_2 $(\mathrm{V^{1/2}\ m^{-1/2}Pa^{-1/2}})$	D_1 $(\mathrm{cm^{-1}\,torr^{-1}})$	D_2 $(\mathrm{V^{1/2}\ cm^{-1/2}torr^{-1/2}})$
He	3.3	12.1	4.4	14.0
Ne	6.2	14.7	8.2	17.0
Ar	21.92	23.01	29.22	26.64
Kr	26.76	24.43	35.69	28.21
Xe	48.98	31.25	65.30	36.08

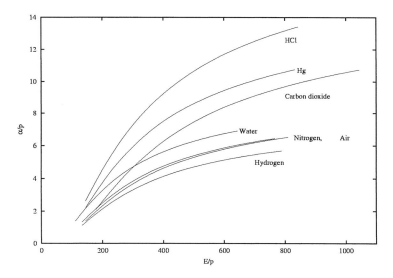

Fig. 3.1 Curves of α_{T}/p (in $\mathrm{m^{-1}Pa^{-1}}$) as a function of E/p (in $\mathrm{V\,m^{-1}Pa^{-1}}$) for various different gases, using the formula (3.61) with parameters derived from Table 3.2. The data for noble gases are shown in Figures 3.2 and 3.3.

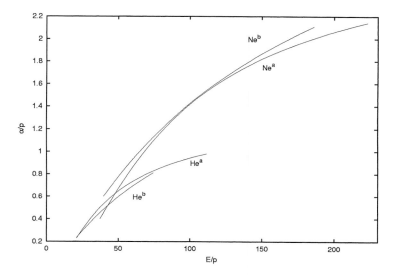

Fig. 3.2   Curves of α_T/p (in $\mathrm{m^{-1}Pa^{-1}}$) as a function of E/p (in $\mathrm{V\,m^{-1}Pa^{-1}}$) for He and Ne, comparing (a) the Townsend formula (3.61), using Table 3.2 with (b) the empirical formula (3.62) using the coefficients in Table 3.3 and the validity ranges quoted in [94].

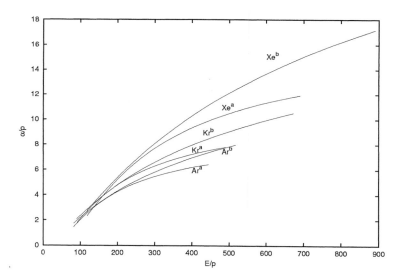

Fig. 3.3   Curves of α_T/p (in $\mathrm{m^{-1}Pa^{-1}}$) as a function of E/p (in $\mathrm{V\,m^{-1}Pa^{-1}}$) for Ar, Kr and Xe, comparing (a) the Townsend formula (3.61), using Table 3.2 with (b) the empirical formula (3.62) using the coefficients in Table 3.3 and the validity ranges quoted in [94].

which yields

$$\frac{\alpha_\mathrm{T}}{p} = \frac{E}{p}\mathcal{F}'\left(\frac{E}{p}\right) \tag{3.65}$$

Equation (3.65) defines the *Stoletow point* for a gas [45], which corresponds to the point on a curve of α_T/p versus E/p at which the tangent to the curve passes through zero.

When the formula (3.61) is used, the Stoletow point occurs at $E/p = C_2$. The Stoletow point is the minimum of the Paschen breakdown curve for a gas (see Section 3.5.5).

3.5.1.3 Restrictions Note that $\alpha_\mathrm{T}/p = \mathcal{F}(E/p)$ is only valid if no pressure dependent ionization processes are operative [64]

3.5.2 Alfvén Ionization

A neutral gas in relative motion with respect to a magnetised plasma will be quickly ionized if the relative speed exceeds the Alfvén critical speed v_c, given by [7, 55, 65]

$$v_c = \left(\frac{2e\phi_i}{m_n}\right)^{\frac{1}{2}} \tag{3.66}$$

where ϕ_i and m_n are respectively the ionization potential and mass of the neutral gas particles, and the plasma and neutral gas have the same chemical composition. The plasma is assumed to be held by the magnetic field, with the flow of neutral atoms producing collisions with the plasma ions. The ions are displaced from their equilibrium positions, producing a significant charge imbalance which cannot be rectified rapidly because of the magnetic field inhibiting electron transport. Hence a local sheath is formed which ionizes the advancing neutral gas very efficiently. Such an effect has been used experimentally to ionize neutral gases in cylindrical geometry via azimuthally driven plasmas [34], and generalized to describe astrophysical flows and shocks [65]. Further details are given in Section 8.5.

3.5.3 Secondary Electron Emission

As the electrode separation d increases whilst maintaining a uniform electric field E, discharge currents greater than that predicted by (3.56) occur. This is attributed to the creation of additional charged particles over and above those generated by primary Townsend ionization.

3.5.3.1 Townsend's Second Ionization Coefficient Townsend's model for this process involves a second ionization coefficient γ_T to account for the secondary emission of electrons by positive ion bombardment of the cathode, leading to a greater electron population, enhanced ionization, and therefore a larger anode current (neglecting diffusion losses)[45, 59, 60, 64, 79]

$$i_a = i_0 \frac{e^{\alpha_\mathrm{T} d}}{1 - \gamma_\mathrm{T}(e^{\alpha_\mathrm{T} d} - 1)} \tag{3.67}$$

The current is increased by the factor $(1 - \gamma_\mathrm{T}(e^{\alpha_\mathrm{T} d} - 1))^{-1}$. Note that the enhanced current can be the result of processes other than positive ion bombardment of the cathode; for example, photoemission at the cathode from excited atoms will yield additional electrons, but without the accompanying positive ions. The electron multiplication described by γ_T and (3.67) applies to all secondary emission effects, although it is possible to distinguish between these processes in a generalised treatment (see Sections 3.5.3.2 and 3.5.3.3). If the secondary electrons are produced solely by positive ion bombardment of the cathode, then γ_T is the number of secondary electrons produced per incident positive ion. Note that γ_T is a function of the electrode composition.

3.5.3.2 Effect of Electron Attachment

An electron colliding with a neutral atom can produce a negative ion in a process termed electron attachment. Since a collision of this type does not produce a further electron via ionization, then it must reduce the ionization rate in the discharge. This can be accounted for in modifying the Townsend model, by defining β_e to be the number of attachments per electron per unit length of drift, in analogy with α_T. Then the effective first ionization coefficient is $\alpha_\mathrm{T} - \beta_e$, with the anode current now given by [45]

$$i_a = i_0 \frac{\alpha_\mathrm{T} e^{(\alpha_\mathrm{T} - \beta_e)d} - \beta_e}{\alpha_\mathrm{T} - \beta_e - \alpha_\mathrm{T} \gamma_\mathrm{T}\{e^{(\alpha_\mathrm{T} - \beta_e)d} - 1\}} \tag{3.68}$$

3.5.3.3 Generalized Treatment of Secondary Processes

A generalized model of anode current i_a produced as a result of a range of secondary effects is given by [60]

$$i_a = \frac{i_0(1 - \kappa/\alpha_\mathrm{T})e^{(\alpha_\mathrm{T} - \kappa)d}}{[1 - \kappa(1 - \delta d)/\alpha_\mathrm{T} - (e^{(\alpha_\mathrm{T} - \kappa)d} - 1)(\gamma_\mathrm{T} + \delta/(\alpha_\mathrm{T} - \kappa) + \kappa(1 - \delta d)/\alpha_\mathrm{T})]} \tag{3.69}$$

where:
α_T is Townsend's first ionization coefficient;
γ_T is Townsend's second ionization coefficient;
$\delta n_e \mathrm{d}x$ is the number of photoelectrons emitted from the cathode as a result of the photons produced by n_e electrons travelling a distance $\mathrm{d}x$ along the electric field;
$\kappa n_i \mathrm{d}x$ is the number of electrons produced collisionally by n_i positive ions travelling a distance $\mathrm{d}x$ along the electric field.

3.5.4   Townsend Breakdown Criterion

The Townsend current (3.67) becomes infinite when

$$\gamma_{\mathrm{T}} e^{\alpha_{\mathrm{T}} d} = \gamma_{\mathrm{T}} + 1 \tag{3.70}$$

known as the breakdown criterion, or the sparking criterion. When satisfied, (3.70) defines the condition where the number of secondary electrons produced by $e^{\alpha_{\mathrm{T}} d}$ positive ions or photons exceeds (by unity) the number of electrons emitted from the cathode as a result of a single ion from a single primary electron. Thus (3.70) is the transition to a self-sustaining discharge, one which is independent of the original ionization source.

Extending the result to include electron attachment yields

$$\gamma_{\mathrm{T}} = \frac{\alpha_{\mathrm{T}} - \beta_e}{\alpha_{\mathrm{T}} (e^{(\alpha_{\mathrm{T}} - \beta_e) d} - 1)} \tag{3.71}$$

so that γ_{T} has to be higher for breakdown (or d, α_{T} greater for fixed γ_{T}).

3.5.5   Paschen Curve

Using the Townsend breakdown condition (3.70) together with the Townsend primary ionization model (3.61) results in the relation [79]

$$C_1 pd \exp\left(-\frac{C_2 pd}{V_b}\right) = \ln\left(1 + \frac{1}{\gamma_{\mathrm{T}}}\right) \tag{3.72}$$

where the breakdown voltage V_b is given by

$$V_b = E_b d \tag{3.73}$$

in plane geometry. Hence

$$V_b = \frac{C_2 pd}{\ln\left[\dfrac{C_1 pd}{\ln(1 + 1/\gamma_{\mathrm{T}})}\right]} \tag{3.74}$$

$$= V_b(pd) \tag{3.75}$$

which is a statement of Paschen's law, that is, the breakdown voltage of a gas depends only on pd (since C_1, C_2 and γ_{T} are fixed for each gas; p and d describe the experimental method).

Note that the minimum breakdown potential, $V_{b,min}$ for a gas occurs at a critical value of pd:

$$(pd)_c = \frac{2.718}{C_1} \ln\left(1 + \frac{1}{\gamma_{\mathrm{T}}}\right) \tag{3.76}$$

$$V_b((pd)_c) = V_{b,min} \leq V_b(pd) \tag{3.77}$$

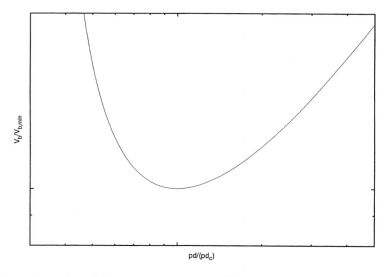

Fig. 3.4 The universal Paschen curve $Y = X/(1 + \log X)$ for the non-dimensional variables $Y = V_b/V_{b,min}$ and $X = (pd)/(pd)_c$

$V_{b,min}$ is also known as the minimum sparking potential.

The functional dependence of V_b on pd can be represented in a universal Paschen curve, defined by [79]

$$\frac{V_b}{V_{b,min}} = \frac{X}{1 + \ln X} \tag{3.78}$$

$$X = \frac{pd}{(pd)_c} \tag{3.79}$$

and shown in Figure 3.4. This curve has a characteristic minimum at $X = 1$, corresponding to the Stoletow point (3.65).

3.6 IONIZATION EQUILIBRIUM

3.6.1 Local Thermodynamic Equilibrium

A gas is in thermal equilibrium if the gas particles are distributed across all possible states according to Boltzmann statistics, and the radiation energy density corresponding to all transitions is given by the black-body curve for the system temperature.

A gas is in local thermodynamic equilibrium (LTE) if it is sufficiently dense for collisional transitions to dominate radiative transitions between all quantum states of the gas particles. This means that the distribution of states follows Boltzmann statistics, but the radiation from such an ensemble of states is not necessarily thermal [46].

3.6.2 Saha Equation

Denoting the number density of atoms in the jth ionization state by n_j as a result of interacting with a co-existing population n_e electrons, then the relative population in the different ionization states in LTE is given by the Saha equation:

$$\frac{n_e n_{j+1}}{n_j} = \frac{2}{h^3} \frac{g_{j+1}}{g_j} (2\pi m_e k_B T)^{3/2} e^{-\xi_j/(k_B T)} \tag{3.80}$$

where g_j is the degeneracy of the jth excited state, ξ_j is the energy difference between the states j and $j+1$, and T is the ensemble temperature.

4
Radiation

4.1 NOTATION

Symbol	Meaning	Ref
$\boldsymbol{A}$	magnetic vector potential	
$\boldsymbol{B}$	magnetic flux density	
c	speed of light in vacuo	
$\boldsymbol{E}$	electric field	
$\boldsymbol{E}_i$	incident electric field	
$\boldsymbol{E}_s$	scattered electric field	
$\boldsymbol{H}$	magnetic intensity	
J_m	Bessel function of 1st kind, order m	
$\boldsymbol{k}$	scattering wave vector	
$\boldsymbol{k}_i$	wave vector of incident electromagnetic wave	
$\boldsymbol{k}_s$	wave vector of scattered electromagnetic wave	
k_B	Boltzmann constant	
K_m	modified Bessel function, order m	
m_s	mass of particle of species s	
n_s	number density of particles of species s	
P	power	
q, q_s	charge on a particle (of species s)	
$\boldsymbol{r}$	position vector from origin to field point	(4.1)
$\boldsymbol{r}_0$	position vector from origin to source point	(4.1)
r_e	classical electron radius	(4.72)
$\boldsymbol{R}$	position vector from source to field point	(4.1)
$\hat{\boldsymbol{R}}$	unit vector in $\boldsymbol{R}$ direction	
s	label defining species: i (ion), e (electron), n (neutral)	
T_e	electron temperature	
α	normalised wavenumber, $= k\lambda_D$	(4.64)
$\boldsymbol{\beta}_v$	normalised particle velocity, $= \boldsymbol{v}/c$	
$\boldsymbol{\gamma}_v$	relativistic factor, $= (1 - \beta_v^2)^{-1/2}$	
ϵ	ratio of photon energy to scatterer energy, $= \hbar\omega_i/(mc^2)$	
ϵ_0	vacuum permittivity	
λ_D	debye length	(2.17)
μ_0	vacuum permeability	
$\boldsymbol{\Pi}$	polarization operator	(4.67)
σ_e	Thomson scattering cross-section for single electron	(4.69)
σ_{KN}	Klein-Nishina scattering cross-section	(4.100)
σ_{sc}	scattering cross-section	(2.29)
ω	frequency of electromagnetic wave	
ω_{ce0}	circular cyclotron frequency of rest electron	
Ω	solid angle	

4.2 RADIATION FROM A MOVING POINT CHARGE

The point charge, value q, has position and velocity $\boldsymbol{r}_0$ and $\boldsymbol{v}_0 = \dot{\boldsymbol{r}}_0$ respectively; $\boldsymbol{r}_0$ is termed the source point. Observations of the field pattern arising from the point charge dynamics are made at the stationary field point $\boldsymbol{r}$. The radius vector from the source to the field point is denoted $\boldsymbol{R}$, defined by

$$\boldsymbol{R} = \boldsymbol{r} - \boldsymbol{r}_0 \qquad (4.1)$$

with the appropriate unit vector $\hat{\boldsymbol{R}} = \boldsymbol{R}/R$. A detailed analysis of the treatment of the radiation field from an accelerated charge can be found in [12, 15, 24, 46, 47, 93].

4.2.1 Liénard-Wiechert Potentials

The Liénard-Wiechert potentials describe the radiation field from a moving point charge. The electromagnetic field produced at the field point $\boldsymbol{r}$ at time t by the particle motion can be derived from the Liénard-Wiechert potentials,

$$\phi(\boldsymbol{r}, t) = \frac{q}{4\pi\epsilon_0} \left[\frac{1}{\kappa R} \right]_{ret} \quad \text{(SI)} \qquad (4.2)$$

$$= q \left[\frac{1}{\kappa R} \right]_{ret} \quad \text{(cgs)} \qquad (4.3)$$

$$\boldsymbol{A}(\boldsymbol{r}, t) = \frac{\mu_0 q}{4\pi} \left[\frac{\boldsymbol{v}_0}{\kappa R} \right]_{ret} \quad \text{(SI)} \qquad (4.4)$$

$$= \frac{q}{c} \left[\frac{\boldsymbol{v}_0}{\kappa R} \right]_{ret} \quad \text{(cgs)} \qquad (4.5)$$

where ϕ is the electric potential, $\boldsymbol{A}$ is the magnetic vector potential, $\kappa = 1 - \boldsymbol{R} \cdot \boldsymbol{v}_0/(cR)$ and the notation $[\cdots]_{ret}$ denotes that the expression within the brackets has to be evaluated at the retarded time t', where

$$t' = t - [R]_{ret}/c = t - R(t')/c \qquad (4.6)$$

since the field pattern detected at time t is generated by the charge dynamics at an earlier, retarded time, allowing for the electromagnetic disturbance to propagate.

4.2.2 Electric and Magnetic Fields of a Moving Charge

The electric field arising from the charged particle motion can be written in the form [15, 24, 93]

$$\boldsymbol{E} = \frac{q}{4\pi\epsilon_0} \left[\frac{(\hat{\boldsymbol{R}} - \boldsymbol{\beta}_v)(1 - \beta_v^2)}{\kappa^3 R^2} + \frac{\hat{\boldsymbol{R}} \times \{(\hat{\boldsymbol{R}} - \boldsymbol{\beta}_v) \times \dot{\boldsymbol{\beta}}_v\}}{c\kappa^3 R} \right]_{ret} \quad \text{(SI)} \qquad (4.7)$$

$$= q \left[\frac{(\hat{\boldsymbol{R}} - \boldsymbol{\beta}_v)(1 - \beta_v^2)}{\kappa^3 R^2} + \frac{\hat{\boldsymbol{R}} \times \{(\hat{\boldsymbol{R}} - \boldsymbol{\beta}_v) \times \dot{\boldsymbol{\beta}}_v\}}{c\kappa^3 R} \right]_{ret} \quad \text{(cgs)} \qquad (4.8)$$

The corresponding magnetic field $\boldsymbol{H}$ is given by

$$\boldsymbol{H} = \frac{cq}{4\pi} \left[\frac{(1 - \beta_v^2)\boldsymbol{\beta}_v}{\kappa^3 R^2} + \frac{\kappa\dot{\boldsymbol{\beta}}_v + (\hat{\boldsymbol{R}} \cdot \dot{\boldsymbol{\beta}}_v)\boldsymbol{\beta}_v}{c\kappa^3 R} \right]_{ret} \times \left[\hat{\boldsymbol{R}} \right]_{ret} \quad \text{(SI)} \qquad (4.9)$$

$$= q \left[\frac{(1 - \beta_v^2)\boldsymbol{\beta}_v}{\kappa^3 R^2} + \frac{\kappa\dot{\boldsymbol{\beta}}_v + (\hat{\boldsymbol{R}} \cdot \dot{\boldsymbol{\beta}}_v)\boldsymbol{\beta}_v}{c\kappa^3 R} \right]_{ret} \times \left[\hat{\boldsymbol{R}} \right]_{ret} \quad \text{(cgs)} \qquad (4.10)$$

Note that the expressions for the electric and magnetic field consist of two terms: a *near field* term, proportional to $[1/R^2]_{ret}$ and a *radiation* term, proportional to $[1/R]_{ret}$. The near field term is essentially the instantaneous Coulomb field of the point charge. If the acceleration is zero, so that the charge is moving uniformly in a straight line, then only the near field term exists, and we have

$$\boldsymbol{E} = \frac{q}{4\pi\epsilon_0}(1 - \beta_v^2) \left[\frac{\hat{\boldsymbol{R}} - \boldsymbol{\beta}_v}{\kappa^3 R^2} \right]_{ret} \quad \text{(SI)} \qquad (4.11)$$

$$= q(1 - \beta_v^2) \left[\frac{\hat{\boldsymbol{R}} - \boldsymbol{\beta}_v}{\kappa^3 R^2} \right]_{ret} \quad \text{(cgs)} \qquad (4.12)$$

$$\boldsymbol{H} = \frac{cq}{4\pi}(1 - \beta_v^2) \left[\frac{\boldsymbol{\beta}_v \times \hat{\boldsymbol{R}}}{\kappa^3 R^2} \right]_{ret} \quad \text{(SI)} \qquad (4.13)$$

$$= q(1 - \beta_v^2) \left[\frac{\boldsymbol{\beta}_v \times \hat{\boldsymbol{R}}}{\kappa^3 R^2} \right]_{ret} \quad \text{(cgs)} \qquad (4.14)$$

4.2.3 Power Radiated by an Accelerating Point Charge

Considering only the radiation term, the power $dP(t')/d\Omega$ radiated per unit solid angle at the source point (that is, at the position of the charge) is given by

$$\frac{dP(t')}{d\Omega} = \frac{q^2}{16\pi^2\epsilon_0 c\kappa^5} \left| \hat{\boldsymbol{R}} \times [(\hat{\boldsymbol{R}} - \boldsymbol{\beta}_v) \times \dot{\boldsymbol{\beta}}_v] \right|^2 \quad \text{(SI)} \qquad (4.15)$$

$$= \frac{q^2}{4\pi c\kappa^5} \left| \hat{\boldsymbol{R}} \times [(\hat{\boldsymbol{R}} - \boldsymbol{\beta}_v) \times \dot{\boldsymbol{\beta}}_v] \right|^2 \quad \text{(cgs)} \qquad (4.16)$$

4.2.3.1 Non-Relativistic Where $\beta_v \ll 1$, and so $\kappa \to 1$,

$$\frac{dP(t')}{d\Omega} \approx \frac{q^2\dot{v}_0^2}{16\pi^2\epsilon_0 c^3} \sin^2\theta \quad \text{(SI)} \qquad (4.17)$$

$$\approx \frac{q^2\dot{v}_0^2}{4\pi c^3} \sin^2\theta \quad \text{(cgs)} \qquad (4.18)$$

where θ is the angle between $\dot{v}_0$ and $\hat{R}$. Integrating over all angles yields the Larmor formula for the total power P radiated by a classical particle,

$$P = \frac{q^2 \dot{v}_0^2}{6\pi\epsilon_0 c^3} \quad \text{(SI)} \tag{4.19}$$

$$= \frac{2q^2 \dot{v}_0^2}{3c^3} \quad \text{(cgs)} \tag{4.20}$$

The radiation pattern is shown in Figure 4.1.

4.2.3.2 Relativistic, $\boldsymbol{\beta_v}$, $\dot{\boldsymbol{\beta}}_v$ collinear

Where the velocity and acceleration are collinear,

$$\frac{dP(t')}{d\Omega} = \frac{q^2 \dot{\beta}_v^2}{16\pi^2\epsilon_0 c} \frac{\sin^2\theta}{(1 - \beta_v \cos\theta)^5} \quad \text{(SI)} \tag{4.21}$$

$$= \frac{q^2 \dot{\beta}_v^2}{4\pi c} \frac{\sin^2\theta}{(1 - \beta_v \cos\theta)^5} \quad \text{(cgs)} \tag{4.22}$$

where θ is the angle between the velocity vector and the position vector of the field point. The total power radiated is given by

$$P = \frac{q^2}{6\pi\epsilon_0 c} \frac{\dot{\beta}_v^{\,2}}{(1 - \beta_v^2)^3} \quad \text{(SI)} \tag{4.23}$$

$$= \frac{2q^2}{3c} \frac{\dot{\beta}_v^{\,2}}{(1 - \beta_v^2)^3} \quad \text{(cgs)} \tag{4.24}$$

The radiation pattern is shown in Figure 4.2, for the case $\beta_v = 0.4$. Notice that the pattern is distorted towards the forward direction, with the radiation cone having an angular width of $\sim 1/\gamma_v$.

4.2.3.3 Relativistic, $\boldsymbol{\beta_v}$, $\dot{\boldsymbol{\beta}}_v$ orthogonal

The particular case of acceleration orthogonal to velocity is relevant to the motion around a magnetic field line. Taking θ to be the angle between the instantaneous $\boldsymbol{\beta}_v$ and $\hat{R}$ projected onto the orbital plane defined by $\boldsymbol{\beta}_v$ and $\dot{\boldsymbol{\beta}}_v$ (as before), and ϕ the angle between $\hat{R}$ and the orbital plane, then the pattern of radiation is given by [49]

$$\frac{dP}{d\Omega} = \frac{q^2 \dot{\beta}_v^2}{16\pi^2\epsilon_0 c} \frac{1}{(1 - \beta_v \cos\theta)^3} \left[1 - \frac{1 - \beta_v^2}{(1 - \beta_v \cos\theta)^2} \sin^2\theta \cos^2\phi \right] \quad \text{(SI)} \tag{4.25}$$

$$= \frac{q^2 \dot{\beta}_v^2}{4\pi c} \frac{1}{(1 - \beta_v \cos\theta)^3} \left[1 - \frac{1 - \beta_v^2}{(1 - \beta_v \cos\theta)^2} \sin^2\theta \cos^2\phi \right] \quad \text{(cgs)} \tag{4.26}$$

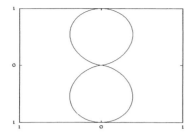

Fig. 4.1 The radiation pattern for a non-relativistic point charge located at (0,0). The left-hand plot shows the cross-section of the pattern in the plane containing the velocity vector and the position vector of the observer. A full 3-D representation of the radiation field is shown on the right.

The total power radiated is then

$$P = \frac{q^2}{6\pi\epsilon_0 c} \frac{\dot{\beta_v}^2}{(1 - \beta_v^2)^2} \quad \text{(SI)} \tag{4.27}$$

$$= \frac{2q^2}{3c} \frac{\dot{\beta_v}^2}{(1 - \beta_v^2)^2} \quad \text{(cgs)} \tag{4.28}$$

The radiation pattern for a point particle with $\beta_v = 0.7$ is shown in Figure 4.3. Once again radiation is beamed in the direction of β_v, with the opening angle of the radiation cone $\sim 1/\gamma_v$. Note the subsidiary radiation maximum at an angle to the main one; as $\beta_v \to 1$ this additional node becomes less important. The radiation pattern does not go to zero along the velocity vector, as in the collinear case. For more detailed discussion of the geometry, see [49].

4.2.3.4 Relativistic, $\boldsymbol{\beta_v}$, $\boldsymbol{\dot{\beta}_v}$ general
The general case has a simple formula for the total power radiated.

$$P = \frac{q^2}{6\pi\epsilon_0 c} \frac{1}{(1 - \beta_v^2)^2} \left(\dot{\beta_v}^2 + \frac{(\boldsymbol{\beta_v} \cdot \boldsymbol{\dot{\beta}_v})^2}{1 - \beta_v^2} \right) \quad \text{(SI)} \tag{4.29}$$

$$= \frac{2q^2}{3c} \frac{1}{(1 - \beta_v^2)^2} \left(\dot{\beta_v}^2 + \frac{(\boldsymbol{\beta_v} \cdot \boldsymbol{\dot{\beta}_v})^2}{1 - \beta_v^2} \right) \quad \text{(cgs)} \tag{4.30}$$

Detailed pictures of the general radiation field can be found in [93].

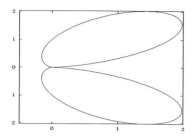

Fig. 4.2 The radiation pattern for a $\beta_v = 0.4$ point charge located at (0,0), in the case of collinear velocity and acceleration. The left-hand plot shows the cross-section of the pattern in the plane containing the velocity vector and the position vector of the observer. Notice that the radiation pattern is now swept forward, pointing in the direction of the particle motion, which is from left to right along the horizontal axis. A full 3-D representation of the radiation field is shown on the right.

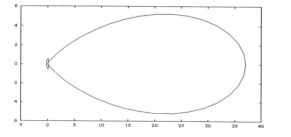

Fig. 4.3 The radiation pattern for a $\beta_v = 0.7$ point charge located at (0,0), in the case of orthogonal velocity and acceleration. The left-hand plot shows the cross-section of the pattern in the orbital plane containing the velocity vector and the acceleration vector, where the velocity is directed along the horizontal axis, pointing towards the right. Notice that the radiation pattern is once more swept forward, pointing in the direction of the particle motion but without a zero in the direction of motion. Note also the small subsidiary maximum. A full 3-D representation of the radiation field is shown on the right, viewed from slightly behind the particle to show the subsidiary maximum.

4.2.4   Frequency Spectrum of Radiation from an Accelerating Charge

The energy distribution in frequency space per unit solid angle, $\mathrm{d}W/\mathrm{d}\Omega$ of the radiation from an accelerated point charge is expressed as

$$\frac{\mathrm{d}W}{\mathrm{d}\omega} = \int_0^\infty \frac{\mathrm{d}^2 W}{\mathrm{d}\Omega\,\mathrm{d}\omega}\,\mathrm{d}\omega \qquad (4.31)$$

where

$$\frac{\mathrm{d}^2 W(\omega)}{\mathrm{d}\Omega\,\mathrm{d}\omega} = \frac{\omega^2 q^2}{16\pi^2\epsilon_0 c} \times$$

$$\left| \int_{-\infty}^\infty \exp\left[i\omega(t' - \hat{\boldsymbol{R}} \cdot \boldsymbol{r}_0(t')/c) \right] \left[\hat{\boldsymbol{R}} \times \left(\hat{\boldsymbol{R}} \times \boldsymbol{\beta}_v \right) \right] \mathrm{d}t' \right|^2 \quad \text{(SI)} \qquad (4.32)$$

$$= \frac{\omega^2 q^2}{4\pi c} \times$$

$$\left| \int_{-\infty}^\infty \exp\left[i\omega(t' - \hat{\boldsymbol{R}} \cdot \boldsymbol{r}_0(t')/c) \right] \left[\hat{\boldsymbol{R}} \times \left(\hat{\boldsymbol{R}} \times \boldsymbol{\beta}_v \right) \right] \mathrm{d}t' \right|^2 \quad \text{(cgs)} \qquad (4.33)$$

4.3   CYCLOTRON AND SYNCHROTRON RADIATION

An electron undergoing cyclotron or Larmor orbits at a source point will produce electromagnetic radiation at the distant field point. This radiation will appear at characteristic frequencies determined by the kinetic energy of the electron, and the magnitude of the magnetic field at the source. Qualitatively, the nature of the radiation may be classified as follows:

low energy, 'classical' electron	line emission at the fundamental electron cyclotron frequency
moderate energy relativistic electron	harmonics of the fundamental occur within an emission envelope
high energy, ultra relativistic electron	smooth continuum emission across a wide frequency range, termed synchrotron emission

Assume that the uniform magnetic field of magnitude B lies along the z-axis, and that the distant observer's field point lies in the x, z-plane, so that

$$\hat{\boldsymbol{R}} = \hat{\mathbf{x}}\,\sin\theta + \hat{\mathbf{z}}\,\cos\theta \qquad (4.34)$$

Take the electron's position and velocity vectors to be

$$\boldsymbol{r}_0(t') = \frac{c\beta_\perp}{\omega_{ce}} \left[\hat{\mathbf{x}} \cos(\omega_{ce}t') + \hat{\mathbf{y}} \sin(\omega_{ce}t') \right] + \hat{\mathbf{z}}\,\beta_\parallel t' \tag{4.35}$$

$$\boldsymbol{\beta}_v = \beta_\perp \left[\hat{\mathbf{x}} \cos(\omega_{ce}t') + \hat{\mathbf{y}} \sin(\omega_{ce}t') \right] + \hat{\mathbf{z}}\,\beta_\parallel \tag{4.36}$$

$$= \dot{\boldsymbol{r}}_0$$

where ω_{ce} is the cyclotron frequency for electrons, given by

$$\omega_{ce} = eB/(\gamma_v m_{e0}) \quad \text{(SI)} \tag{4.37}$$
$$= eB/(\gamma_v m_{e0}c) \quad \text{(cgs)} \tag{4.38}$$
$$= \omega_{ce0}/\gamma_v \tag{4.39}$$

and where $\beta_v = v/c$, $\gamma_v = (1 - \beta_v^2)^{-1/2}$ are the usual relativistic parameters.

Note that (4.35) and (4.36) assume that the electron orbit is unaffected by radiation losses.

Radiation at the field point is detected at frequencies $\omega = \omega_m$ given by [15, 24, 46, 47]

$$\omega_m = \frac{m\omega_{ce}}{1 - \beta_\parallel \cos\theta} \tag{4.40}$$

$$= \frac{(1 - \beta_\perp^2 - \beta_\parallel^2)^{1/2}}{1 - \beta_\parallel \cos\theta} m\omega_{ce0} \tag{4.41}$$

where $m = 1, 2, \ldots$.

Restrictions

- Unless otherwise stated, it is assumed that there is no radiation-reaction on the particle, that is, the particle's trajectory is unaffected by radiation losses

- the accelerating magnetic field is homogeneous

- only the radiation term is used in the calculations

- the unit vector $\hat{\boldsymbol{R}}$ from the source to the field point is taken to be independent of time, so that although the electron is moving with respect to the observer, the effect of that motion on their relative orientation is negligible

4.3.1 Spectral Power Density

The spectral power density, that is, the energy radiated per unit time per unit frequency, at the field point is given by the Schott-Trubnikov formula

[15, 46, 47]:

$$\frac{\mathrm{d}^2 P}{\mathrm{d}\omega\mathrm{d}\Omega} = \frac{e^2\omega^2}{8\pi^2\epsilon_0 c} \sum_{m=1}^{\infty} \left[\left(\frac{\cos\theta - \beta_\parallel}{\sin\theta} \right)^2 J_m^2(\xi) + \beta_\perp^2 J_m'^2(\xi) \right]$$

$$\times \frac{\delta[(1 - \beta_\parallel \cos\theta)\omega - m\omega_{ce}]}{1 - \beta_\parallel \cos\theta} \quad \text{(SI)} \qquad (4.42)$$

$$= \frac{e^2\omega^2}{2\pi c} \sum_{m=1}^{\infty} \left[\left(\frac{\cos\theta - \beta_\parallel}{\sin\theta} \right)^2 J_m^2(\xi) + \beta_\perp^2 J_m'^2(\xi) \right]$$

$$\times \frac{\delta[(1 - \beta_\parallel \cos\theta)\omega - m\omega_{ce}]}{1 - \beta_\parallel \cos\theta} \quad \text{(cgs)} \qquad (4.43)$$

where δ is the delta-function: $\delta(0) = 1$, $\delta(x) = 0$ for all $x \neq 0$; J_m is the Bessel function of the first kind, of order m; J_m' is the derivative of the Bessel function with respect to its argument; and

$$\xi = \frac{\omega\beta_\perp \sin\theta}{\omega_{ce}} \qquad (4.44)$$

Note that (4.42) refers to the radiation detected at the field point; (4.42) needs to be multiplied by $1 - \beta_\parallel \cos\theta$ in order to find the energy loss at the particle per unit solid angle per unit frequency.

4.3.2   Power in Each Harmonic

For each harmonic m, the radiated power P_m detected at the field point is given by

$$P_m = \frac{e^2\omega_{ce0}^2}{2\pi\epsilon_0\gamma_v^2\beta_\perp(1 - \beta_\parallel^2)^{3/2}}$$

$$\times \left[m\beta_\perp^2 J_{2m}'(\zeta) - \frac{m^2}{\gamma_v^2} \int_0^{\beta_\perp/(1-\beta_\parallel^2)^{1/2}} J_{2m}(2mt)\mathrm{d}t \right] \quad \text{(SI)} \qquad (4.45)$$

$$= \frac{2e^2\omega_{ce0}^2}{c\gamma_v^2\beta_\perp(1 - \beta_\parallel^2)^{3/2}}$$

$$\times \left[m\beta_\perp^2 J_{2m}'(\zeta) - \frac{m^2}{\gamma_v^2} \int_0^{\beta_\perp/(1-\beta_\parallel^2)^{1/2}} J_{2m}(2mt)\mathrm{d}t \right] \quad \text{(cgs)} \qquad (4.46)$$

where

$$\zeta = \frac{2m\beta_\perp}{(1 - \beta_\parallel^2)^{1/2}} \qquad (4.47)$$

4.3.3 Total Radiated Power

The total radiated power P_{total} at the field point due to the power in each harmonic P_m is given by [15, 46, 47]

$$P_{total} = \frac{e^2 \omega_{ce0}^2}{6\pi\epsilon_0 c} \frac{\beta_\perp^2}{1 - \beta_v^2} \quad \text{(SI)} \tag{4.48}$$

$$= \frac{2e^2 \omega_{ce0}^2}{3c} \frac{\beta_\perp^2}{1 - \beta_v^2} \quad \text{(cgs)} \tag{4.49}$$

4.3.4 $\beta_v \ll 1$: Cyclotron Emission

In the non-relativistic limit, $P_{m+1}/P_m \sim \beta_v^2$ and the bulk of the emission is concentrated in the fundamental, the cyclotron emission line, giving the power detected at the field point per unit solid angle as

$$\frac{\mathrm{d}P}{\mathrm{d}\Omega} \simeq \frac{e^2 \omega_{ce0}^2 \beta_\perp^2}{32\pi^2 \epsilon_0 c} \beta_\perp^2 (1 + \cos^2 \theta) \quad \text{(SI)} \tag{4.50}$$

$$\simeq \frac{e^2 \omega_{ce0}^2 \beta_\perp^2}{8\pi c} \beta_\perp^2 (1 + \cos^2 \theta) \quad \text{(cgs)} \tag{4.51}$$

showing that the power detected when the observer is aligned with the magnetic field direction is twice that detected in the orthogonal orientation. This is because in the former, the electron motion is circular, and the resultant radiation contains both linear polarizations. Observations made at 90° to the magnetic field can only see one of the linear polarizations, and therefore detect only half the power.

4.3.5 $\beta_v \sim 1$: Synchrotron Emission

For ultra-relativistic particles, the emission spectrum ceases to be discrete lines but instead becomes a smooth continuum. It is more appropriate therefore to consider the total power per unit frequency interval,

$$\frac{\mathrm{d}P(\omega)}{\mathrm{d}\omega} = \sqrt{3} \frac{e^2 \omega_{ce0}}{8\pi^2 \epsilon_0 c} \frac{\omega}{\omega^*} \int_{\omega/\omega^*}^\infty K_{5/3}(x)\mathrm{d}x \quad \text{(SI)} \tag{4.52}$$

$$= \sqrt{3} \frac{e^2 \omega_{ce0}}{2\pi c} \frac{\omega}{\omega^*} \int_{\omega/\omega^*}^\infty K_{5/3}(x)\mathrm{d}x \quad \text{(cgs)} \tag{4.53}$$

where

$$\omega^* = \tfrac{3}{2}\gamma_v^2 \omega_{ce0} \tag{4.54}$$

The term ω^* can be expressed in terms of the local radius of curvature R_c of the relativistic electron [47]:

$$\omega^* = \frac{3}{2}\gamma^3 c/R_c \qquad (4.55)$$

The total energy per unit frequency $I(\omega)$ radiated by the ultra-relativistic electron can then be expressed as

$$I(\omega) \simeq \sqrt{3}\,\frac{e^2\gamma\omega}{4\pi\epsilon_0 c\omega^*}\int_{\omega/\omega^*}^{\infty} K_{5/3}(x)\mathrm{d}x \qquad \text{(SI)} \qquad (4.56)$$

$$\approx \frac{\sqrt{3}}{2}\,\frac{e^2\gamma}{4\pi\epsilon_0 c}\left(\frac{\omega}{\omega^*}\right)^{1/2}\exp(-\omega/\omega^*), \quad \omega \gg \omega^* \quad \text{(SI)} \qquad (4.57)$$

$$\simeq \sqrt{3}\frac{e^2\gamma\omega}{c\omega^*}\int_{\omega/\omega^*}^{\infty} K_{5/3}(x)\mathrm{d}x \qquad \text{(cgs)} \qquad (4.58)$$

$$\approx \frac{\sqrt{3}}{2}\frac{e^2\gamma}{c}\left(\frac{\omega}{\omega^*}\right)^{1/2}\exp(-\omega/\omega^*), \quad \omega \gg \omega^* \quad \text{(cgs)} \qquad (4.59)$$

4.4   BREMSSTRAHLUNG

Plasma electrons moving in the electric field of plasma ions will also radiate, due to the accompanying accelerations produced by unshielded ion fields.

The power P_e radiated by a single electron moving in the field of a stationary ion, the latter carrying charge q_i, is

$$P_e = \frac{q_i^2 e^4}{96\pi^3\epsilon_0^3 c^3 m_e^2 r_{ei}^4} \qquad \text{(SI)} \qquad (4.60)$$

$$= \frac{2q_i^2 e^4}{3c^3 m_e^2 r_{ei}^4} \qquad \text{(cgs)} \qquad (4.61)$$

where r_{ei} is the electron-ion separation distance, and m_e the electron mass. Integrating (4.60) over all electron encounters with this same ion, assuming uniform electron number density, and then generalising to account for all ions, yields the classical result

$$P_{tot} = \frac{q_i^2 e^4 n_i n_e}{24\pi^2\epsilon_0^3 c^3 m_e \hbar}\left(\frac{k_B T_e}{m_e}\right)^{\frac{1}{2}} \qquad \text{(SI)} \qquad (4.62)$$

$$= \frac{8\pi q_i^2 e^4 n_i n_e}{3c^3 m_e \hbar}\left(\frac{k_B T_e}{m_e}\right)^{\frac{1}{2}} \qquad \text{(cgs)} \qquad (4.63)$$

where the singularity is removed by a minimum cut-off $r_{min} \sim \hbar(m_e k_B T_e)^{1/2}$ taken as the de Broglie wave number.

Restrictions   Note that (4.62) is restricted to electron-ion collisions, and is a classical calculation; the full quantum mechanical treatment yields a result which is numerically very close.

4.5   RADIATION SCATTERING

A plane monochromatic electromagnetic wave incident on a free electron at rest will accelerate it, producing radiation from the accelerated particle. There are different theoretical treatments of the scattered radiation field, depending on the particle or plasma parameter regimes, characterised by the following quantities:

$$
\begin{aligned}
&\omega_i && \text{frequency of incident wave} \\
&\omega_s && \text{frequency of detected scattered wave} \\
&\omega = \omega_s - \omega_i && \text{`scattering' frequency} \\
&\boldsymbol{k}_i && \text{wave-vector of incident wave} \\
&\boldsymbol{k}_s && \text{wave-vector of detected scattered wave} \\
&\boldsymbol{k} = \boldsymbol{k}_s - \boldsymbol{k}_i && \text{`scattering' wave-vector} \\
&\alpha = k\lambda_D && \\
&\boldsymbol{\beta}_v = \boldsymbol{v}/c && \text{normalised particle velocity} \\
&\epsilon = \frac{\hbar\omega_i}{m_e c^2} && \text{ratio of photon energy to scatterer energy}
\end{aligned}
\tag{4.64}
$$

The following table shows which theoretical treatment is appropriate for parameter ranges.

α	β	ϵ	Scattering Description	Section
-	$\ll 1$	$\ll 1$	single particle, non-relativistic Thomson	4.69
$\gg 1$	$\ll 1$	$\ll 1$	incoherent, non-relativistic Thomson	4.5.2.1
$\gg 1$	< 1	$\ll 1$	incoherent, relativistic Thomson	4.5.2.2
$\ggg 1$	< 1	$\ll 1$	coherent, relativistic Thomson	4.5.3
-	< 1	$\nless 1$	single particle, relativistic, Compton	4.5.4
-	< 1	$\nless 1$	single particle, relativistic, quantum Klein-Nishina	4.5.5

If the incident photons have negligible energies compared to the rest mass of the scatterer (that is $\hbar\omega \ll m_s c^2$), this process is termed *Thomson Scattering*, and the scattering particle trajectory can be prescribed under the influence of the incident wave without accounting for the effect of the radiation on that trajectory [46, 51, 82]

For higher energy photons, the recoil of the scatterer must be accounted for; the correct dynamics are described by the *Compton Scattering Process*. The cross-section for the scattering of high-energy photons is the *Klein-Nishina Cross-Section*, which correctly accounts for the relativistic and quantum-mechanical aspects of the electron-photon interaction. The Klein-Nishina formula reproduces the Thomson result at low photon energies.

4.5.1   Thomson Scattering

Here it is assumed in all circumstances that the incident photons have negligible energy compared with the rest energy of the electron: $\hbar\omega \ll m_e c^2$.

The detected radiation field at the distant observer can be evaluated under different approximations, depending on whether the scattering is coherent ($k\lambda_D < 1$) or incoherent ($k\lambda_D \gg 1$).

For $k\lambda_D \gg 1$, that is, incoherent scattering, there are two sub-cases: (i) non-relativistic, where the dipole approximation, using the Larmor formula (4.19) is adequate; and (ii) relativistic, where the full relativistic form of the radiation pattern is needed, usually for electrons at a temperature of $\sim 1\,\mathrm{eV}$ and above.

For $k\lambda_D < 1$, that is, coherent scattering, a coherent sum must be formed of the electric fields radiated by the participating electrons, leading to the requirement for an accurate scattering form factor for coherent scattering from discrete sites.

Following [46], the scattered electric field $\boldsymbol{E}_s$ is given by

$$\boldsymbol{E}_s = r_e \Big[R^{-1}(1-\beta_v^2)^{\frac{1}{2}}(1-\beta_s)^{-3} \times \{$$

$$- \hat{e}(1-\beta_{vi})(1-\beta_{vs})$$

$$- \hat{\boldsymbol{i}}\beta_{ve}(1-\beta_{vs})$$

$$+ \hat{s}([1-\beta_{vi}]\hat{s}\cdot\hat{e} + [\hat{s}\cdot\hat{\boldsymbol{i}} - \beta_{vs}]\beta_{ve})$$

$$- \boldsymbol{\beta}_v([1-\beta_{vi}]\hat{s}\cdot\hat{e} - [1 - \hat{s}\cdot\hat{\boldsymbol{i}}]\beta_{ve})\}\Big]_{ret}$$

where: $\hat{\boldsymbol{i}} = \boldsymbol{k}/k$ is the unit vector in the propagation direction of the incident wave; $\hat{e} = \boldsymbol{E}_i/E_i$ is the unit vector in the direction of the incident electric field; $\hat{s} = \hat{\boldsymbol{R}}$ is the unit vector in the direction of the scattered wave; and

$$\beta_{vi} = \boldsymbol{\beta}_v \cdot \hat{\boldsymbol{i}} \tag{4.65}$$

$$\beta_{vs} = \boldsymbol{\beta}_v \cdot \hat{s} \tag{4.66}$$

The scattering geometry is shown in Figure 4.4.

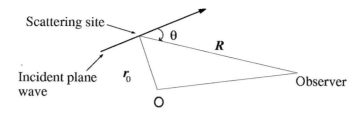

Fig. 4.4 Diagram showing the directions of the incident electromagnetic wave and the scattered wave

For notational convenience, it is possible to write (4.65) in a compact form,

$$\boldsymbol{E}_s = \left[\frac{r_e}{R}\boldsymbol{\Pi} \cdot \boldsymbol{E}_i\right]_{ret} \tag{4.67}$$

where $\boldsymbol{\Pi}$ is a polarization operator for the radiation field, given by (4.7).

4.5.1.1 Thomson Scattering Cross-Section for single electron

The scattering cross-section σ_e for non-relativistic single electron radiation using only the Larmor formula (4.19) for the radiated power, for which

$$\boldsymbol{\Pi} \cdot \boldsymbol{E}_i = \hat{\boldsymbol{s}} \times (\hat{\boldsymbol{s}} \times \boldsymbol{E}_i) \tag{4.68}$$

yields [46, 50]

$$\sigma_e = \frac{8\pi}{3} \left(\frac{e^2}{4\pi\epsilon_0 m_e c^2}\right)^2 \quad \text{(SI)} \tag{4.69}$$

$$= \frac{8\pi}{3} \left(\frac{e^2}{m_e c^2}\right)^2 \quad \text{(cgs)} \tag{4.70}$$

$$= \frac{8\pi}{3} r_e^2 \tag{4.71}$$

where r_e is the classical electron radius, defined by

$$r_e = \frac{e^2}{4\pi\epsilon_0 m_e c^2} \quad \text{(SI)} \tag{4.72}$$

$$= \frac{e^2}{m_e c^2} \quad \text{(cgs)} \tag{4.73}$$

The differential Thomson scattering cross-section for a single isolated electron is given by

$$\frac{\mathrm{d}\sigma_e}{\mathrm{d}\Omega} = \tfrac{1}{2}r_e^2(1 + \cos^2\theta) \tag{4.74}$$

where θ is the angle between the direction of the incident wave and scattered wave; see Figure 4.4. An alternative form is

$$\frac{\mathrm{d}\sigma_e}{\mathrm{d}\Omega} = r_e^2 \sin^2 \phi \tag{4.75}$$

where ϕ is the angle between $\hat{e}$ and $\hat{s}$.

If the incident wave is monochromatic with frequency ω_i and wavevector $\boldsymbol{k}_i$, then the scattered field from the single electron also has a single frequency ω_s given by

$$\omega_s = \omega_i + c(\boldsymbol{k}_s - \boldsymbol{k}_i) \cdot \boldsymbol{\beta}_v \tag{4.76}$$

$$= \omega_i \frac{1 - \hat{i} \cdot \boldsymbol{\beta}_v}{1 - \hat{s} \cdot \boldsymbol{\beta}_v} \tag{4.77}$$

where $c\beta_v$ is the electron's velocity.

Restrictions

- $\hbar\omega_i \ll m_e c^2$, so that negligible net momentum is imparted to the electron by the radiation field

- strictly, (4.71) and (4.74) apply only to single, non-relativistic electrons subject to an electromagnetic wave of frequency ω such that $\hbar\omega \ll m_e c^2$

- there is no collective plasma effect

- the Larmor formula (4.19) is applicable

4.5.2 Incoherent Thomson Scattering from an Unmagnetized Plasma

4.5.2.1 Non-Relativistic Plasma, $k\lambda_D \gg 1$ In the limit $k\lambda_D \gg 1$ where k is the wavenumber of the incident radiation then the phases at each scattering site in the plasma can be taken as random, and the angular pattern of total fraction of scattered radiation is simply the sum of the individual contributions:

$$\frac{\mathrm{d}\sigma_{sc}}{\mathrm{d}\Omega} = n_e \tfrac{1}{2} r_e^2 (1 + \cos^2 \theta) \tag{4.78}$$

Restrictions

- $k\lambda_D \gg 1$ ensuring that the 'scattering' wavelength is sufficiently small that no coherence effects are present

- $\hbar\omega_i \ll m_e c^2$ so that negligible net momentum is imparted to the electron from the radiation

- $\omega_i, \omega_s \gg \omega_p, \omega_{ce}$ so that the incident and scattered frequencies are able to propagate without being modified by plasma effects

- $E_i \ll E_D$, where E_D is the Dreicer field, the threshold for the electron-runaway instability (9.112)

- the plasma electron motion is non-relativistic

For a distribution f of electrons under the non-relativistic (dipole) approximation,

$$\frac{\mathrm{d}^2\sigma}{\mathrm{d}\Omega\mathrm{d}\omega_s} = \left[r_e^2 \int_V \langle \boldsymbol{S}_i \rangle \mathrm{d}\boldsymbol{r} |\hat{\boldsymbol{s}} \times (\hat{\boldsymbol{s}} \times \hat{\boldsymbol{e}})|^2 f_k(\omega/k)/k \right] / \left(\int_V \langle \boldsymbol{S}_i \rangle n_e \mathrm{d}\boldsymbol{r} \right) \quad (4.79)$$

$$= \sigma_e \sin^2(\phi) f_k(\omega/k)/(n_e k), \quad (4.80)$$

where

$$f_k(v_k) = \int \mathrm{d}\boldsymbol{v}_\perp f(\boldsymbol{v}_\perp, v_k) \quad (4.81)$$

$$= n_e \left(\frac{m_e}{2\pi k_B T_e} \right)^{1/2} \exp\left(-\frac{m_e v_k^2}{2k_B T_e} \right) \quad (4.82)$$

and where V is the scattering volume, $\omega = \omega_s - \omega_i$ and $\boldsymbol{k} = \boldsymbol{k}_s - \boldsymbol{k}_i$, $\langle \boldsymbol{S}_i \rangle$ is the mean incident Poynting vector, and (4.82) holds for a Maxwellian electron plasma. Note that (4.80) and (4.82) assume that the electron number density is constant in the scattering volume.

4.5.2.2 Relativistic Plasma, $k\lambda_D \gg 1$

For electrons with energies in excess of 1 eV, relativistic terms become important and full expression for the radiation term (4.65) must be used. The result here for the differential scattering cross-section per unit frequency is

$$\frac{\mathrm{d}^2\sigma_p}{\mathrm{d}\Omega\mathrm{d}\omega_s} = r_e^2 \frac{\omega_i}{\omega_s} \int |\Pi \cdot \hat{\boldsymbol{e}}|^2 \frac{\kappa^2}{n_e} f\delta(\boldsymbol{k} \cdot \boldsymbol{v} - \omega)\mathrm{d}\boldsymbol{v} \quad (4.83)$$

where $\boldsymbol{k} = \boldsymbol{k}_s - \boldsymbol{k}_i$ and $\kappa = 1 - \hat{\boldsymbol{s}} \cdot \boldsymbol{\beta}_v$. Note that (4.83) assumes that the electron number density is constant in the scattering volume; the general case requires the volume integration over the mean incident Poynting vector times the number density as a normalisation (see (4.79)).

Restrictions The same restrictions apply as in the previous section, except that here the prescription of the radiation field is not predicated on the Larmor or dipole approximation.

For the simpler case in which E_i is perpendicular to both $\hat{s}$ and $\hat{i}$, the scattered power can be written as

$$\frac{\mathrm{d}^2 P}{\mathrm{d}\Omega \mathrm{d}\omega_s} = r_e^2 \int_V \langle S_i \rangle \mathrm{d}r \left| \frac{1 - \beta_{vi}}{1 - \beta_{vs}} \right|^2 \times$$

$$\int \left| 1 - \frac{(1 - \hat{s} \cdot \hat{i})\beta_{ve}^2}{(1 - \beta_{vi})(1 - \beta_{vs})} \right|^2 (1 - \beta_v^2) f(v_\perp, v_k)/k \mathrm{d}v_\perp \qquad (4.84)$$

There are various approximations to (4.84) appropriate to particular circumstances; see [71, 81] for details.

4.5.3   Coherent Thomson Scattering from an Unmagnetized Plasma

For $k\lambda_D < 1$ there is significant correlation between electrons necessitating a coherent sum of scattering contributions from each particle. The total scattered power spectrum can be expressed in the form

$$\frac{\mathrm{d}^2 P}{\mathrm{d}\omega \mathrm{d}\Omega} = \frac{r_e^2}{2\pi} \frac{P_i}{A} |\mathbf{\Pi} \cdot \hat{e}|^2 n_e V S_f(\mathbf{k}, \omega) \qquad (4.85)$$

where P_i is the total incident power delivered to the volume V, A is the area of V perpendicular to $\mathbf{k}$, n_e is the electron number density in V, and $S_f(\mathbf{k}, \omega)$ is the scattering form factor, which can be approximated as [46, 80]

$$S_f(\mathbf{k}, \omega) \approx \frac{\sqrt{\pi}}{kv_{th,e}} \Gamma_a(\xi_e) + \frac{\sqrt{\pi}}{kv_{th,i}} (1 + \alpha^2)^{-1} \Gamma_b(\xi_i) \qquad (4.86)$$

$$a = 1/\alpha \qquad (4.87)$$

$$\alpha = k\lambda_D \qquad (4.88)$$

$$b = Z(1 + \alpha^2)^{-1} \frac{T_e}{T_i} \qquad (4.89)$$

$$\xi_s = \frac{\omega}{\sqrt{2}kv_{th,s}} \qquad (4.90)$$

$$v_{th,s} = \left(\frac{k_B T_s}{m_s} \right)^{1/2} \qquad (4.91)$$

$$\Gamma_y(x) = \frac{e^{-x^2}}{[1 + y^2(1 - f(x))]^2 + \pi y^2 x^2 e^{-2x^2}} \qquad (4.92)$$

$$f(x) = 2xe^{-x^2} \int_0^x e^{t^2} \mathrm{d}t \qquad (4.93)$$

Note that this approximation is only valid for one species of ion, carrying charge Ze. Coherent scattering will arise from electron plasma waves, and

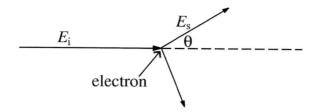

Fig. 4.5 Diagram showing the directions of the incident photon, the scattered photon and the electron recoil for a Compton scattering event.

ion-acoustic waves. The Salpeter approximation (4.86) to the form factor simplifies the full expression to account for these two effects; note that it breaks down for very large T_e/T_i.

4.5.4 Compton Scattering

For $\hbar\omega_i \not\ll m_e c^2$ then the incident radiation affects the evolution of the electron trajectory, with consequences for the scattered radiation field. From Figure 4.5, energy and momentum conservation yields [49]

$$\frac{1}{\hbar\omega_s} + \frac{1}{\hbar\omega_i} = \frac{1 - \cos\theta}{m_e c^2} \qquad (4.94)$$

or equivalently,

$$\lambda_s - \lambda_i = \frac{h}{m_e c}(1 - \cos\theta) \qquad (4.95)$$

where λ_i, λ_s are the wavelengths of the incident and scattered radiation respectively. The Compton shift is the quantity $\lambda_s - \lambda_i$. The Compton shift for 90° scattering is known as the Compton wavelength of the electron λ_c:

$$\lambda_c = \frac{h}{m_e c} = 2.426 \times 10^{-12}\,\mathrm{m} \qquad (4.96)$$

4.5.5 Klein-Nishina Cross-Section

The cross-section for Compton Scattering demands a relativistic, quantum mechanical treatment of the electron interaction with the high-energy photon: the Klein-Nishina cross-section. For highly energetic photons, the Thomson cross-section for a single electron at rest has to be modified for the scattering of unpolarized incident radiation. The differential cross-section is given by [76, 99]

$$\frac{\mathrm{d}\sigma_{KN}}{\mathrm{d}\Omega} = \tfrac{1}{2}r_e^2 \frac{\omega_s^2}{\omega_i^2} \left(\frac{\omega_i}{\omega_s} + \frac{\omega_s}{\omega_i} - \sin^2\theta \right) \tag{4.97}$$

$$= \tfrac{1}{2}r_e^2 \left[\frac{1}{1 + \epsilon(1 - \cos\theta)} \right]^3 (1 + \cos^2\theta) \times$$

$$\left[1 + \frac{\epsilon^2(1 - \cos\theta)^2}{(1 + \cos^2\theta)\{1 + \epsilon(1 - \cos\theta)\}} \right] \tag{4.98}$$

where

$$\epsilon = \frac{\hbar\omega_i}{m_e c^2} \tag{4.99}$$

The total cross-section is then [96, 99]

$$\sigma_{KN} = \frac{\pi r_e^2}{\epsilon} \left\{ \left[1 - \frac{2(\epsilon + 1)}{\epsilon^2} \right] \ln(2\epsilon + 1) + \frac{1}{2} + \frac{4}{\epsilon} - \frac{1}{2(2\epsilon + 1)^2} \right\} \tag{4.100}$$

$$\approx \sigma_e (1 - 2\epsilon + 5.2\epsilon^2 - 13.3\epsilon^3 + 32.7\epsilon^4 - 77.7\epsilon^5 + \cdots), \quad (\epsilon < \tfrac{1}{2}) \tag{4.101}$$

$$\sim \frac{\pi r_e^2}{\epsilon} (\ln(2\epsilon) + \tfrac{1}{2}) \quad \text{for } \epsilon \gg 1 \tag{4.102}$$

Note that σ_{KN} is the same as the Thomson cross section for $\epsilon \ll 1$, and also for $\epsilon \not\ll 1$, $\theta = 0$, but falls away rapidly with θ for high energy photons.

5

Kinetic Theory

5.1 NOTATION

SYMBOL	MEANING	REF
B	magnetic field	
E	electric field	
f	distribution function	
f_0	equilibrium distribution function	
f_D	Druyvesteyn distribution function	(5.47)
f_M	Maxwell-Boltzmann distribution function	(5.9)
F_M	f_M expressed as a distribution of speeds	(5.10)
g	energy distribution function	(5.6)
g_p	energy probability function	(5.7)
$\mathcal{G}$	Fokker-Planck potential	(5.36)
$\mathcal{H}$	Fokker-Planck potential	(5.37)
J_{ext}	external current density	(5.20),(5.22)
k_B	Boltzmann constant	
m_e	electron mass	
m_n	neutral particle mass	
n	particle number density	(5.1)
$\mathbf{P}$	pressure tensor	(5.3)
q	heat flux vector	(5.5)
r	position vector	
T_g	gas (neutral) temperature	
u	particle velocity	
$\bar{u}$	bulk velocity	(5.2)
Γ	gamma function	
ε	energy density	(5.4)
Λ	argument in Coulomb logarithm	(6.15)
$\lambda_{\mathrm{mfp},e}$	mean free path for electrons	(2.19)
$\lambda_{\mathrm{mfp},g}$	mean free path for gas particles	(2.19)
ν_c	electron-neutral collision frequency	
ξ	energy loss factor	(5.40)
ρ_{ext}	external charge density	(5.19),(5.21)

5.2 FUNDAMENTALS

The distribution function f [19, 85] is the statistical [44] description of the plasma or gas particles in the six-dimensional space (r, u) at a given time t. Thus $f(r, u, t)\mathrm{d}r\mathrm{d}u$ is the number of particles at time t having velocities in the range $u \rightarrow u + \mathrm{d}u$ in the infinitesimal spatial volume $r \rightarrow r + \mathrm{d}r$.

Bulk or fluid quantities are derived from the distribution function via moments of f in velocity space:

$$n(\mathbf{r}, t) = \int f(\mathbf{r}, \mathbf{u}, t) \mathrm{d}\mathbf{u} \qquad \text{number density} \qquad (5.1)$$

$$\bar{\mathbf{u}}(\mathbf{r}, t) = \frac{1}{n} \int \mathbf{u} f \mathrm{d}\mathbf{u} \qquad \text{bulk velocity} \qquad (5.2)$$

$$\mathbf{P}(\mathbf{r}, t) = m \int (\mathbf{u} - \bar{\mathbf{u}})(\mathbf{u} - \bar{\mathbf{u}}) f \mathrm{d}\mathbf{u} \qquad \text{pressure tensor} \qquad (5.3)$$

$$\varepsilon = \frac{m}{2n} \int (\mathbf{u} - \bar{\mathbf{u}})^2 f \mathrm{d}\mathbf{u} \qquad \text{energy density} \qquad (5.4)$$

$$\mathbf{q} = \frac{m}{2} \int (\mathbf{u} - \bar{\mathbf{u}})^2 (\mathbf{u} - \bar{\mathbf{u}}) f \mathrm{d}\mathbf{u} \quad \text{heat flux} \qquad (5.5)$$

Two transformations of the general distribution function are also widely used, particularly in experimental research. These are: the energy distribution function (EDF) $g(\epsilon)$, defined by

$$g(\epsilon)\mathrm{d}\epsilon = 4\pi u^2 f(u) \mathrm{d}u \qquad (5.6)$$

where $\epsilon = mu^2/2$; and the energy probability function (EPF) $g_p(\epsilon)$, defined by

$$g_p(\epsilon) = \epsilon^{-1/2} g(\epsilon). \qquad (5.7)$$

5.3 BOLTZMANN EQUATION

The fundamental equation governing the evolution of the distribution function is the Boltzmann equation:

$$\frac{\partial f}{\partial t} + \mathbf{u} \cdot \frac{\partial f}{\partial \mathbf{r}} + \mathbf{a} \cdot \frac{\partial f}{\partial \mathbf{u}} = \left(\frac{\partial f}{\partial t} \right)_c \qquad (5.8)$$

where $\mathbf{a}$ is the acceleration experienced by the particles, and the term on the right is the rate of change with respect to time of the distribution function in response to collisions. This collision term is the key one which defines the basic physics content in any modelling.

5.4 MAXWELLIAN DISTRIBUTION

The Boltzmann equation (5.8) has as a unique equilibrium solution, in the absence of external fields, the Maxwell-Boltzmann distribution function, also known as the Maxwellian:

$$f_M(\boldsymbol{u}) = n \left(\frac{m}{2\pi k_B T}\right)^{3/2} \exp\left(-\frac{m(\boldsymbol{u}-\bar{\boldsymbol{u}})^2}{2k_B T}\right) \tag{5.9}$$

A useful variant of (5.9) is one which describes the distribution of speeds, rather than vector velocities. Taking a non-streaming plasma, and integrating (5.9) over all possible directions yields

$$F_M(u) = 4\pi n \left(\frac{m}{2\pi k_B T}\right)^{3/2} u^2 \exp\left(-\frac{mu^2}{2\pi k_B T}\right) \tag{5.10}$$

Using (5.10), two important quantities are associated with the Maxwellian distribution: the most probable speed of a particle, corresponding to the maximum of the distribution function, and defined by

$$u_p = \left(\frac{2k_B T}{m}\right)^{1/2} \tag{5.11}$$

and the average speed,

$$\bar{u} = \left(\frac{8k_B T}{\pi m}\right)^{1/2} \tag{5.12}$$

The root-mean-square speed u_{rms} is

$$u_{\mathrm{rms}} = \left(\frac{3k_B T}{m}\right)^{1/2} \tag{5.13}$$

Note that the EDF for a Maxwellian takes the form

$$g(\epsilon) = \frac{2n}{(k_B T)^{3/2}} \epsilon^{1/2} \exp\left(-\frac{\epsilon}{k_B T}\right) \tag{5.14}$$

5.4.0.1   Restrictions on the Maxwellian Distribution   For a Maxwellian velocity distribution to be a good approximation, there are restrictions on the spatial dependence of the gas temperature T_g and number density n, and also on the magnitude of any electric field $\boldsymbol{E}$ which may be present. These are:

$$\lambda_{cg}\frac{\partial T_g}{\partial r} \ll T_g \tag{5.15}$$

$$\lambda_{cg}\frac{\partial n}{\partial r} \ll n \tag{5.16}$$

$$Ee\lambda_c \ll \tfrac{1}{2}mu^2 \tag{5.17}$$

where λ_{cg}, λ_c are the gas and electron mean free paths, respectively (see Section 2.19).

5.5   VLASOV DESCRIPTION

The simplest form of (5.8) is where the explicit collision term is set to zero, and the Lorentz force per unit mass is taken as the acceleration. The averaged long-range nature of the collective plasma behaviour is incorporated by demanding that the electromagnetic field terms are those arising self-consistently from the distribution of plasma particles. The whole system is called the Vlasov description of a fully ionised plasma:

$$\frac{\partial f}{\partial t} + \boldsymbol{u} \cdot \frac{\partial f}{\partial \boldsymbol{r}} + \boldsymbol{a} \cdot \frac{\partial f}{\partial \boldsymbol{u}} = 0 \tag{5.18}$$

$$\sum_s \frac{q_s}{\epsilon_0} \int f_s(\boldsymbol{r}, \boldsymbol{u}, t) \mathrm{d}\boldsymbol{u} + \frac{\rho_{\text{ext}}}{\epsilon_0} = \nabla \cdot \boldsymbol{E}(\boldsymbol{r}, t) \tag{5.19}$$

$$\mu_0 \sum_s q_s \int \boldsymbol{u} f_s(\boldsymbol{r}, \boldsymbol{u}, t) \mathrm{d}\boldsymbol{u} + \epsilon_0 \mu_0 \frac{\partial \boldsymbol{E}}{\partial t} + \mu_0 \boldsymbol{J}_{\text{ext}} = \nabla \times \boldsymbol{B}(\boldsymbol{r}, t) \tag{5.20}$$

Equations (5.19) and (5.20) are in SI units; the equivalent forms in cgs units are:

$$\sum_s 4\pi q_s \int f_s(\boldsymbol{r}, \boldsymbol{u}, t) \mathrm{d}\boldsymbol{u} + 4\pi \rho_{\text{ext}} = \nabla \cdot \boldsymbol{E}(\boldsymbol{r}, t) \tag{5.21}$$

$$\frac{4\pi}{c} \sum_s q_s \int \boldsymbol{u} f_s(\boldsymbol{r}, \boldsymbol{u}, t) \mathrm{d}\boldsymbol{u} + \frac{1}{c} \frac{\partial \boldsymbol{E}}{\partial t} + \frac{4\pi}{c} \boldsymbol{J}_{\text{ext}} = \nabla \times \boldsymbol{B}(\boldsymbol{r}, t) \tag{5.22}$$

$\boldsymbol{J}_{ext}$ and ρ_{ext} represent the externally supplied current and charge density respectively.

5.5.1   Equilibrium Solutions

Note that the Vlasov equation has many equilibrium solutions f_{s0} which satisfy $\partial f_{s0}/\partial t = 0$. In the field-free case, the primary equilibrium solution is the Maxwellian, given by (5.9). However, the neglect of an explicit collision term in the Vlasov construction admits metastable equilibrium solutions, that is, solutions which are stable on a timescale comparable with the collision time, and which ultimately will relax to a Maxwellian. Such equilibria can be written in general as arbitrary functions of the constants of the motion of a charged particle in the electric and magnetic fields.

5.5.1.1   Case I: $\boldsymbol{E} = \boldsymbol{B} = 0$ Here the constants of the motion are the energy $\epsilon = mu^2/2$ and the momentum $\boldsymbol{p} = m\boldsymbol{u}$. Hence any function $f_0 =$

$f_0(u_x, u_y, u_z)$ is a metastable equilibrium function. Examples include [50]:

$$f_0 = \frac{u_0}{2(u^4 + u_0^4)} \qquad (5.23)$$

$$f_0 = u_0 \delta(u_x) \delta(u_y) \delta(u_z^2 - u_0^2) \qquad (5.24)$$

$$f_0 = \left(\frac{m}{2\pi k_B T} \right)^{1/2} \delta(u_x) \delta(u_y) \exp \left(-\frac{m(u_z^2 - u_0^2)}{2k_B T} \right) \qquad (5.25)$$

5.5.1.2   Case II: $E = 0$, $B = \hat{z} B_0(r)$ If subscript $\perp$ denotes components in the plane perpendicular to the magnetic field, then a simple equilibrium solution is

$$f_0 = f_0(u_\perp, u_z) \qquad (5.26)$$

As in the previous case, the constants of the motion can also feature in the construction, in particular, the adiabatic invariants associated with orbit theory (see Section 6.4.2).

5.5.1.3   Case III: $E = -\hat{x} \, \partial\phi(x)/\partial x, B = 0$ Constants of the motion here are the y- and z-momenta, and the energy in the x-direction, $mu_x^2/2 + q\phi(x)$. Thus a possible equilibrium solution is

$$f_0 = f_0(u_x^2 + 2q\phi(x)/m, u_y, u_z). \qquad (5.27)$$

5.5.1.4   Stability of Meta-Equilibria   A stable equilibrium is one for which the kinetic energy is a constant. It is sufficient for stability that f_0 be a monotonically decreasing function of u^2, that is,

$$\frac{\partial f_0}{\partial u^2} < 0. \qquad (5.28)$$

5.6   COLLISIONAL MODELLING

The collision term on the right-hand side of (5.8) can be modelled in several different ways, each appropriate for a restricted range of physical significance. Fundamental to all approaches is the Coulomb collision cross Section $\sigma_c(u_0, \theta_c)$, defined in Section 2.5.1.

5.6.1   Boltzmann Collision Term

A distribution f of interacting particles can be modelled by considering the reciprocal communication between particles in the assembly to be fundamentally binary in nature. Hence the distribution function evolves according to binary interactions which scatter a certain particle population out of a particular velocity space element, accompanied by other interactions which scatter different populations into that same velocity space element. Labelling the 'scattered' population with subscript 1, and the 'scattering' population with subscript 2, Boltzmann constructed the collision term

$$\left(\frac{\partial f}{\partial t}\right)_c = \int \left(f'(u_1)f'(u_2) - f(u_1)f(u_2)\right) \mid u_1 - u_2 \mid \sigma_c(\mid u_1 - u_2 \mid, \theta)\mathrm{d}\Omega\mathrm{d}u,$$

(5.29)

where Ω is the solid angle.

5.6.1.1 Restrictions The Boltzmann collision term is strictly only valid if:

- every interaction is a binary one;

- all interactions are uncorrelated;

- each interaction must take place over length scales and time scales much less than any intrinsic variation in f.

5.6.2 Simplified Boltzmann Collision Term

An approximate form of (5.29) is

$$\left(\frac{\partial f}{\partial t}\right)_c = \frac{f_0 - f(r, u, t)}{\tau}$$

(5.30)

where a single time τ between collisions is used to characterise the collisional relaxation from the perturbed distribution f to the equilibrium solution f_0. This form is usually referred to as the Krook collision term.

5.6.2.1 Restrictions

- f_0 should be chosen to conserve particle number, e.g. a local Maxwellian

- the collision operator (5.30) will drive f to a stationary equilibrium, which may not be appropriate if momentum is to be conserved. If f describes the evolution of electrons in the presence of stationary massive scattering particles, then (5.30) is a good approximation. Such a model is termed a Lorentz gas; since there is only self-interaction included in (5.30), it applies best when there are mainly neutral species present.

5.6.3 Fokker-Planck

In order to account for the many weak interactions which characterise a fully ionised plasma, the Fokker-Planck collision term defines a function $\psi(u, \Delta u)$ which describes the probability that a particle with initial velocity u undergoes many small-angle scattering interactions in a time Δt such that it acquires a velocity increment $\Delta\boldsymbol{u}$. Since ψ is independent of time, and therefore the particle's history, the scattering process is Markovian.

The formal statement of the collision term is then

$$\left(\frac{\partial f}{\partial t}\right)_c = -\frac{\partial}{\partial \boldsymbol{u}} \cdot (f\langle\Delta\boldsymbol{u}\rangle) + \frac{1}{2}\frac{\partial^2}{\partial\boldsymbol{u}\partial\boldsymbol{u}} : (f\langle\Delta\boldsymbol{u}\Delta\boldsymbol{u}\rangle) \qquad (5.31)$$

where

$$\langle\Delta\boldsymbol{u}\rangle = \lim_{\Delta t\to 0}\frac{1}{\Delta t}\int \psi\Delta\boldsymbol{u}\mathrm{d}(\Delta\boldsymbol{u}) \qquad (5.32)$$

$$\langle\Delta\boldsymbol{u}\Delta\boldsymbol{u}\rangle = \lim_{\Delta t\to 0}\frac{1}{\Delta t}\int \psi\Delta\boldsymbol{u}\Delta\boldsymbol{u}\mathrm{d}(\Delta\boldsymbol{u}) \qquad (5.33)$$

These two terms represent the two main ways in which an evolving distribution function can change as a result of collisions: the velocity of a group of particles may be changed as a result of many weak interactions, a process termed dynamical friction and described by (5.32); the velocity of a group of particles may be spread about in velocity space, a process termed velocity diffusion, and quantified by (5.33).

5.6.4 Fokker-Planck Potentials

A formulation of the Fokker-Planck collision term using potential functions [78, 98] can be written as follows:

$$\left(\frac{\partial f}{\partial t}\right)_c = \Gamma_p\left[-\frac{\partial}{\partial\boldsymbol{u}}\cdot\left(f\frac{\partial\mathcal{H}}{\partial\boldsymbol{u}}\right) + \frac{1}{2}\frac{\partial^2}{\partial\boldsymbol{u}\partial\boldsymbol{u}} : \left(f\frac{\partial^2\mathcal{G}}{\partial\boldsymbol{u}\partial\boldsymbol{u}}\right)\right], \qquad (5.34)$$

where

$$\Gamma_p = \frac{(q^2 q_{sc}^2)}{4\pi\epsilon_0^2 m^2}\ln\Lambda, \qquad (5.35)$$

m and q are the mass and charge of the test particle; subscript sc denotes the equivalent properties of the scatterer, and the potentials $\mathcal{G}$ and $\mathcal{H}$ are given by

$$\mathcal{G} = \int f_{sc}(\boldsymbol{u}_{sc})\mid\boldsymbol{u} - \boldsymbol{u}_{sc}\mid \mathrm{d}\boldsymbol{u}_{sc} \qquad (5.36)$$

$$\mathcal{H} = \frac{m + m_{sc}}{m}\int\frac{f_{sc}(\boldsymbol{u}_{sc})}{\mid\boldsymbol{u} - \boldsymbol{u}_{sc}\mid}\mathrm{d}\boldsymbol{u}_{sc} \qquad (5.37)$$

5.6.4.1 Restrictions The Fokker-Planck equation (5.31) is valid for any system of particles in which collisions only produce small velocity changes, and for which large velocity changes can only come about from the incremental effect of many small such interactions. Hence this description is valid only if the plasma parameter Γ_p is large, that is if there is a large number of charged particles in a Debye sphere.

5.7 DRIVEN SYSTEMS

Where there is significant time dependence in physical processes other than collisional relaxation, such as electric fields present in the plasma which are driven to change on timescales shorter than relaxation timescales, then these must be accounted for in the kinetic description.

5.7.1 Generalized Distribution

The general form of the electron (speed) distribution function F in a neutral gas, subjected to an oscillating electric field of frequency ω and rms amplitude E_{rms} is [30, 64]:

$$F = A \exp(-W) \tag{5.38}$$

$$W = \int^u mu \left[k_B T_g + \frac{2}{3} \frac{e^2 E_{rms}^2}{m_e (\nu_c^2 + \omega^2) \xi} \right]^{-1} du \tag{5.39}$$

where T_g is the gas temperature, ν is the collision frequency for elastic electron-neutral collisions, and ξ is the energy loss factor [85]:

$$\xi (U_e - U_g) = \quad \text{Average energy transfer per collision} \tag{5.40}$$

where U_e is the electron energy, and U_g is the average energy of the gas molecules. The electron energy gain from collisions with neutral particles can be written as

$$\frac{dU_e}{dt} = -\xi \nu_c (U_e - U_g) \tag{5.41}$$

showing that if $U_g > U_e$ then the electrons are heated by the gas (such as in a shock), and vice-versa.

For elastic collisions in monatomic gases,

$$\xi = \frac{2m_e}{m_e + m_n} \tag{5.42}$$

where m_n is the mass of a gas particle. For molecular gases, ξ has a more complicated form, since contributions from the internal energies of the molecule have to be accounted for. An approximate expression for this latter case in which the neutral particle has transitions between internal energy states is

$$\xi(v) = \left\{ \sum_k \left[\frac{2m_e \nu_{c,k}}{m_n - m_e} + \frac{v_{12,k}^2}{v^2} \frac{m_e v_{12}^2}{2k_B T_g} \nu_k \right] \right\} \left\{ \sum_k \nu_{c,k} \right\}^{-1} \tag{5.43}$$

where the sum is over all possible transitions k, $\nu_{c,k}$ is the collision frequency for the kth transition, v is the electron speed, and $\frac{1}{2} m_e v_{12,k}^2$ is the energy difference between possible transitions.

Three special cases of the general form of the electron speed distribution can be identified:

5.7.1.1 Thermal Motion Dominant: Maxwellian Distribution   Here,

$$k_B T_g \gg \frac{2}{3} \frac{e^2 E_{\mathrm{rms}}^2}{m_e (\nu^2 + \omega^2) \xi} \tag{5.44}$$

and so (5.39) has the solution

$$F = F_M \tag{5.45}$$

5.7.1.2 Thermal Motion Negligible: Druyvesteyn Distribution   Taking $\omega = 0$, $E_{\mathrm{rms}} = E_0$ and $\xi = 2m_e/(m_e + m_n) = $ constant, $\lambda_c = $ constant and assuming negligible thermal energy in the gas, yields

$$W = \frac{3}{2} \xi \left(\frac{\frac{1}{2} m_e u^2}{e E_0 \lambda_c} \right)^2 \tag{5.46}$$

from which the Druyvesteyn distribution function F_D can be defined:

$$F_D = A \exp\left(-B u^4\right) \tag{5.47}$$

$$A = \frac{1}{\pi} \frac{\Gamma^{3/2}(5/4)}{\Gamma^{5/2}(3/4)} \frac{n}{\left(\bar{u^2}\right)^{3/2}} \tag{5.48}$$

$$B = \frac{1}{(\bar{u^2})^2} \left(\frac{\Gamma(5/4)}{\Gamma(3/4)} \right)^2 \tag{5.49}$$

where the average energy is given by

$$\frac{1}{2} m_e \bar{u^2} = \frac{\Gamma(5/4)}{\Gamma(3/4)} \left(\frac{2}{3\xi} \right)^{1/2} e E_0 \lambda_c \tag{5.50}$$

Note that the Druyvesteyn distribution has significantly fewer high-energy electrons in comparison with a Maxwellian for the same total energy content; this is shown in Figure 5.1, in which the Druyvesteyn and Maxwellian distributions are compared.

5.7.1.3 Harmonic E, Thermal Motion Negligible: Amended Druyvesteyn   Assuming $\omega \neq 0$, and ξ and $\lambda_{\mathrm{mfp}} = $ both constant, yields a corrected Druyvesteyn-type distribution:

$$W = \frac{3}{2} \xi \left(\frac{\frac{1}{2} m_e u^2}{e E_{\mathrm{rms}} \lambda_{\mathrm{mfp}}} \right)^2 + \frac{3}{2} \xi \left(\frac{m_e u \omega}{e E_{\mathrm{rms}}} \right)^2 \tag{5.51}$$

5.7.1.4 High Frequency Limit   For the case of $\omega \gg \nu_c$ in (5.51), the distribution function becomes Maxwellian in form, but with an anomalous temperature T^*, which incorporates the energy density of the electric field:

$$W \approx \frac{\frac{1}{2} m u^2}{k_B T^*} \tag{5.52}$$

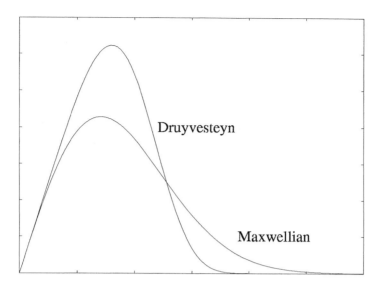

Fig. 5.1 Graphs of the Maxwellian and Druyvesteyn speed distributions for the same total energy content, with particle speed along the horizontal axis. The Druyvesteyn has significantly fewer high-energy particles in comparison with the Maxwellian.

where

$$k_B T^* = k_B T_g + \frac{2eE_{\text{rms}}^2}{3m\omega^2\xi} \tag{5.53}$$

5.7.1.5 General Form   Integrating (5.39) for the general distribution function [64, 85] yields

$$F_0 = C\left(1 + \frac{\epsilon}{\beta_1 + \beta_2}\right)^{\beta_1} \exp(-\epsilon) \tag{5.54}$$

where

$$\epsilon = \frac{mu^2}{2k_B T_g} \tag{5.55}$$

$$\beta_1 = \frac{1}{3}\left(\frac{e^2 E_{\text{rms}}^2 \lambda_{\text{mfp}}^2}{k_B^2 T_g^2 \xi}\right) \tag{5.56}$$

$$\beta_2 = \frac{m\lambda_{\text{mfp}}^2 \omega^2}{2k_B T_g} \tag{5.57}$$

and C is a normalising constant.

5.7.1.6 Restrictions   The construction of (5.39) is valid under the following assumptions:

- the mean free path of electrons is much shorter than the dimensions of the confining vessel;

- the plasma system has reached steady-state;

- the velocity distribution function can be expanded in spherical harmonics, retaining only the first-order terms;

In addition, (5.47), (5.51) and (5.54) are restricted to constant $\xi, \lambda_{\mathrm{mfp}}$.

6

Plasma Transport

6.1 NOTATION

Symbol	Meaning	Ref
b	impact parameter	
b_0	critical impact parameter	(6.6)
$\boldsymbol{B}$	magnetic field	
$\boldsymbol{E}$	electric field	
J_L	longitudinal invariant	(6.52)
k_B	Boltzmann constant	
m	particle mass	
m_e	electron mass	
m_r	reduced mass	(6.1)
n	particle number density	(5.1)
r_L	Larmor radius	(2.21)
T_p	plasma temperature	
T_g	gas (neutral) temperature	
u_r	relative speed	
$W_\parallel$	kinetic energy parallel to $\boldsymbol{B}$	(6.33)
$W_\perp$	kinetic energy perpendicular to $\boldsymbol{B}$	(6.33)
$\boldsymbol{u}_D$	diamagnetic drift velocity	(6.44)
$\bar{\boldsymbol{u}}$	bulk velocity	(5.2)
α_r	test particle scattering factor	(6.20)
β_r	reciprocal thermal speed	(6.22)
Γ	Gamma function	
ε	energy density	(5.4)
Λ	argument in Coulomb logarithm	(6.15)
λ_c	mean free path for electrons	(2.19)
λ_{cg}	mean free path for gas particles	(2.19)
λ_D	Debye length	(2.17)
μ_{bs}	magnetic moment of a particle of species s	(2.33)
ν_c	electron-neutral collision frequency	
σ_{R}	Rutherford differential scattering cross-section	(6.2)
ξ	energy loss factor	(5.40)

6.2 BASIC DEFINITIONS

Elastic Collision: a collision between two bodies in which the total kinetic energy and the total translational momentum before and after the collision are the same.

Inelastic Collision: a collision between two bodies in which the particles before and after collision are the same, but the total kinetic energy has been changed.

These are excitation collisions, in which there is significant energy transfer into internal vibrational or rotational modes of at least one of the bodies.
Transformation Collision: one in which the particles before the collision are different from those after the collision. An example is impact ionisation.

6.3   BINARY COLLISIONS

The asymptotic deflection angle in the centre of mass co-ordinate system for the incident particle relative to its original trajectory is denoted θ_0.

The reduced mass m_r for binary collisions between particles of mass m_1 and m_2 is defined to be

$$m_r = \frac{m_1 m_2}{m_1 + m_2} \tag{6.1}$$

It is essential to use m_r in the centre of mass frame.

6.3.1   Elastic Collisions Between Charged Particles

6.3.1.1   Binary Coulomb Collision   For two particles with charge and mass q_i, m_i, $i = 1, 2$, undergoing a Coulomb interaction at a relative speed u_r and with impact parameter b, the Rutherford differential scattering cross-section for Coulomb collisions is given by

$$\sigma_{\mathrm{R}}(u_r, \theta_0) = \frac{b_0^2}{4 \sin^4(\theta_0/2)} \tag{6.2}$$

where θ_0 is the asymptotic deflection of the incident particle in the centre of mass frame, given by

$$\tan(\theta_0/2) = \frac{b_0}{b} \tag{6.3}$$

$$= \frac{q_1 q_2}{4\pi\epsilon_0 m_r u_r^2 b} \quad \text{(SI)} \tag{6.4}$$

$$= \frac{q_1 q_2}{m_r u_r^2 b} \quad \text{(cgs)} \tag{6.5}$$

and the critical impact parameter b_0 for deflections of 90° is defined as follows:

$$b_0 = \frac{|q_1 q_2|}{4\pi\epsilon_0 m_r u_r^2} \quad \text{(SI)} \tag{6.6}$$

$$= \frac{|q_1 q_2|}{m_r u_r^2} \quad \text{(cgs)} \tag{6.7}$$

The cross-section for scattering through 90° in a single collision is

$$\sigma_{s90°} = \pi b_0^2 \tag{6.8}$$

6.3.1.2 Multiple Coulomb Collisions In a fullly ionised plasma, it is more likely that an electron will suffer many small angle deflections as a result of encounters with ions of charge q. A small angle deflection $\delta\theta$ is characterised by the impact parameter (using (6.3)):

$$\frac{\delta\theta_{max}}{\delta\theta_{min}} \approx \frac{b_{max}}{b_{min}} \tag{6.9}$$

Practical values for the extrema of the impact factor are:

$$b_{max} \approx \lambda_D, \tag{6.10}$$

$$b_{min} \approx \frac{qe}{4\pi\epsilon_0 m_e u_r^2} \approx \frac{qe}{12\pi\epsilon_0 k_B T_p} \quad \text{(SI)} \tag{6.11}$$

$$b_{min} \approx \frac{qe}{m_e u_r^2} \approx \frac{qe}{3k_B T_p} \quad \text{(cgs)} \tag{6.12}$$

Note that the approximation for b_{max} assumes Debye shielding makes deflections negligible for $b > \lambda_D$; the approximation for b_{min} is derived on the basis that only small angle deflections are important, and hence the maximum such angle satisfies $\tan(\theta_{max}/2) \approx \theta_{max}/2 \Rightarrow \theta_{max}/2 \approx 1$.

The cross-section for scattering an electron through 90° as a result of multiple Coulomb collisions with ions is

$$\sigma_{m90°} = 8\pi \left(\frac{qe}{4\pi\epsilon_0 m_e u_r^2}\right)^2 \ln\Lambda \quad \text{(SI)} \tag{6.13}$$

$$\sigma_{m90°} = 8\pi \left(\frac{qe}{m_e u_r^2}\right)^2 \ln\Lambda \quad \text{(cgs)} \tag{6.14}$$

where the term $\ln\Lambda$, referred to as the Coulomb Logarithm, is defined via

$$\Lambda = \frac{b_{max}}{b_{min}}, \tag{6.15}$$

and for most laboratory plasmas, $10 \leq \ln\Lambda \leq 20$.

The ratio of the cross section for multiple scattering through 90°, and single scattering through 90° is given by

$$\frac{\sigma_{m90°}}{\sigma_{s90°}} = 8\ln\Lambda \tag{6.16}$$

showing that a large angle deflection of an electron by multiple Coulomb scattering is at least two orders of magnitude more likely than that arising from a single encounter with an ion.

Restrictions: The results in this section are strictly true only for a Lorentzian gas, that is, a gas of mobile electrons in the presence of stationary ions in which electron-electron interactions are ignored.

6.3.1.3   Relaxation Times for Maxwellian Distributions   A test particle, mass m, charge Z_e travelling at speed u through a Maxwellian distribution of scattering particles, with temperature T_s, charge $Z_s e$, number density n_s and mass m_s, has the following associated timescales for physical processes:

$$\text{slowing down time:} \quad \tau_s = \frac{u}{(1 + m/m_s)\alpha_r \beta_r^2 \Psi(\beta_r u)} \tag{6.17}$$

$$\text{deflection time:} \quad \tau_D = \frac{u}{\alpha_r [\text{erf}(\beta_r u) - \Psi(\beta_r u)]} \tag{6.18}$$

$$\text{energy exchange time:} \quad \tau_E = \frac{u^3}{4\alpha_r \Psi(\beta_r u)} \tag{6.19}$$

where

$$\alpha_r = \frac{Z_s^2 Z^2 e^4 n_s \log \Lambda}{2\pi \epsilon_0^2 m^2} \quad \text{(SI)} \tag{6.20}$$

$$= \frac{8\pi Z_s^2 Z^2 e^4 n_s \log \Lambda}{m^2} \quad \text{(cgs)} \tag{6.21}$$

$$\beta_r = \left(\frac{m_s}{2k_B T_s}\right)^{1/2} \quad \text{(SI \& cgs)} \tag{6.22}$$

$$\Psi(x) = \frac{\text{erf}(x) - x\text{erf}'(x)}{2x^2} \tag{6.23}$$

and where $\text{erf}(x)$ is the error function [2].

Note that τ_s is related to the dynamical friction term in the Fokker-Planck description (5.32); τ_D is the relaxation time for an initially anisotropic distribution to become isotropic, and is derived from the diffusive Fokker-Planck term (5.33); τ_E is the typical time-scale for the relaxation of a homogeneous distribution to Maxwellian form.

Two special cases can be identified: very slow test particles ($\beta u \ll 1$); and very fast test particles ($\beta u \gg 1$). Given that:

$$\text{erf}(x) \rightarrow \begin{cases} 2x/\sqrt{\pi} & \text{as } x \to 0, \\ 1 & \text{as } x \to \infty; \end{cases} \tag{6.24}$$

$$\Psi(x) \rightarrow \begin{cases} 2x/(3\sqrt{\pi}) & \text{as } x \to 0, \\ 1/(2x^2) & \text{as } x \to \infty; \end{cases} \tag{6.25}$$

then

$$
\tau_s \to
\begin{cases}
\dfrac{3\sqrt{\pi}}{2(1+m/m_s)\alpha_r\beta_r^3} & \text{as } \beta_r u \to 0, \\[4mm]
\dfrac{2u^3}{(1+m/m_s)\alpha_r} & \text{as } \beta_r u \to \infty;
\end{cases}
\tag{6.26}
$$

$$
\tau_D \to
\begin{cases}
\dfrac{3\sqrt{\pi}u^2}{4\alpha_r\beta_r} & \text{as } \beta_r u \to 0, \\[4mm]
\dfrac{u^3}{\alpha_r} & \text{as } \beta_r u \to \infty;
\end{cases}
\tag{6.27}
$$

$$
\tau_E \to
\begin{cases}
\dfrac{3\sqrt{\pi}u^2}{8\alpha_r\beta_r} & \text{as } \beta_r u \to 0, \\[4mm]
\dfrac{\beta_r^2 u^5}{2\alpha_r} & \text{as } \beta_r u \to \infty;
\end{cases}
\tag{6.28}
$$

Self-collisions Where the test particle and scatterers are identical, τ_e gives an indication of how long the gas will take to relax to a Maxwellian. Assuming thermal velocities such that $\beta_r u \approx 1$ [15] yields

$$
\tau_E^{self} \approx \frac{(2k_BT/m)}{4\alpha_r\psi(1)}.
\tag{6.29}
$$

Extending the calculation to a distribution of test particles encountering a distribution of scattering (background) particles allows characteristic times to be quantified for two Maxwellian distributions at (slightly) different temperatures to relax to a single distribution, via different processes: electron-electron collisions, τ_{ee}; electron-ion collisions, τ_{ei}; and ion-ion collisions τ_{ii} [19]. Taking a fully ionised plasma of electrons and ions with equilibrium Maxwellian distributions characterised by temperatures T_e and T_i respectively, then

$$
\tau_{ee} \approx \frac{\sqrt{108}\pi\epsilon_0^2(k_BT_e)^{3/2}m_e^{1/2}}{n_e e^4 \log \Lambda}
\tag{6.30}
$$

$$
\tau_{ei} \approx \frac{\sqrt{108}\pi\epsilon_0^2(k_BT_e)^{3/2}m_i}{Z^2 n_i m_e^{1/2} e^4 \log \Lambda}
\tag{6.31}
$$

$$
\tau_{ii} \approx \frac{\sqrt{108}\pi\epsilon_0^2(k_BT_i)^{3/2}m_i^{3/2}}{n_i Z^4 e^4 \log \Lambda}
\tag{6.32}
$$

6.4   PARTICLE DYNAMICS

The following concepts and definitions are relevant throughout this section, which considers particle motion in imposed fields, rather than the collective dynamics of plasmas:

Guiding Centre: The basic motion of a charged particle is gyration in the plane perpendicular to the magnetic field, coupled with translation arising from any electric field or field inhomogeneity. The locus of points defining the centre of the gyration is termed the guiding centre.

Kinetic energy W: The kinetic energy W of a particle with mass m and speed v is defined as

$$W = \frac{1}{2}mv^2$$
$$= \frac{1}{2}mv_\parallel^2 + \frac{1}{2}mv_\perp^2$$
$$= W_\parallel + W_\perp, \qquad (6.33)$$

defining the parallel, and perpendicular, components of the kinetic energy, $W_\parallel$, $W_\perp$, in the directions parallel, and perpendicular, to the direction of the magnetic field.

6.4.1 Drifts

The following drift velocities are for single particle motion in the presence of imposed electric and magnetic fields.

6.4.1.1 Constant E, B A charged particle moving in the constant fields E, B will acquire a drift velocity v_E:

$$v_E = \frac{E \times B}{B^2}, \quad \text{(SI)} \qquad (6.34)$$
$$= c\frac{E \times B}{B^2}. \quad \text{(cgs)} \qquad (6.35)$$

Note that the drift is independent of charge, and so no charge separation arises.

6.4.1.2 Nonuniform E, Uniform B: A spatially non-uniform electric field together with a uniform magnetic field produces a particle drift due to the finite larmor radius effect, modifying (6.34):

$$v_E = \left(1 + \frac{1}{4}r_L^2\nabla^2\right)\frac{E \times B}{B^2} \quad \text{(SI)} \qquad (6.36)$$
$$= c\left(1 + \frac{1}{4}r_L^2\nabla^2\right)\frac{E \times B}{B^2} \quad \text{(cgs)} \qquad (6.37)$$

It is clear that the drift is now different for particles of different mass, and therefore charge separation can occur.

6.4.1.3 Non-Uniform B, E = 0: grad B drift

A particle of charge q moving in the spatially non-uniform magnetic field $\boldsymbol{B}$ will acquire a drift velocity $\boldsymbol{v}_G$

$$\boldsymbol{v}_G = \frac{W_\perp}{qB^3}\left(\boldsymbol{B} \times \nabla\right)B, \quad \text{(SI)} \tag{6.38}$$

$$\boldsymbol{v}_G = c\frac{W_\perp}{qB^3}\left(\boldsymbol{B} \times \nabla\right)B, \quad \text{(cgs)} \tag{6.39}$$

where $W_\perp$ is the particle's kinetic energy in the plane perpendicular to the direction of the magnetic field.

Note that since the drift velocity is charge dependent, this drift can create charge separation.

6.4.1.4 Non-uniform B, E = 0: Curvature Drift

If the magnetic field lines are curved, then not only does the particle of charge q drift according to (6.38), but there is also a curvature drift, given by

$$\boldsymbol{v}_c = \frac{2W_\|}{qB^4}(\boldsymbol{B} \times (\boldsymbol{B} \cdot \nabla)\boldsymbol{B}) \quad \text{(SI)} \tag{6.40}$$

$$= \frac{2cW_\|}{qB^4}(\boldsymbol{B} \times (\boldsymbol{B} \cdot \nabla)\boldsymbol{B}) \quad \text{(cgs)} \tag{6.41}$$

where $W_\|$ is the particle's kinetic energy in the direction of the magnetic field.

6.4.1.5 External Force Drift

The presence of a constant, non-electromagnetic force density $\boldsymbol{f}_{ext}$ in addition to a uniform magnetic field results in a particle drift given by

$$\boldsymbol{v}_f = \frac{m}{qB^2}(\boldsymbol{f}_{ext} \times \boldsymbol{B}) \quad \text{(SI)} \tag{6.42}$$

$$= \frac{mc}{qB^2}(\boldsymbol{f}_{ext} \times \boldsymbol{B}) \quad \text{(cgs)} \tag{6.43}$$

6.4.1.6 Restrictions

The drift velocities above are derived under the following approximations:

- the charged particles do not give rise to any appreciable collective effect (sparse plasma);

- the gradient scale-length for magnetic field variations is much greater than the larmor radius;

- the time-varying fields change on time-scales much less than the cyclotron frequency for the particle concerned;

- radiation is ignored;

- all collisional effects are ignored;

- all speeds are non-relativistic;

6.4.1.7 Uniform B, non-uniform density: Diamagnetic Drift
For a *fluid* plasma satisfying an equation of motion

$$mn\left[\frac{\partial u}{\partial t} + (u \cdot \nabla)u\right] = -\nabla p + qn(E + u \times B) \quad \text{(SI)} \qquad (6.44)$$

$$= -\nabla p + qn(E + u \times B/c) \quad \text{(cgs)} \qquad (6.45)$$

where p is the pressure, and n the plasma number density, the diamagnetic drift velocity u_D is given by

$$u_D = -\frac{\nabla p \times B}{qnB^2} \quad \text{(SI)} \qquad (6.46)$$

$$= -c\frac{\nabla p \times B}{qnB^2} \quad \text{(cgs)} \qquad (6.47)$$

Note that u_D is a fluid, not a particle, concept, in which collisionality is important (see discussions in [23, 98]).

6.4.1.8 Motion in a Monochromatic Plane Wave
A plane polarized electromagnetic wave with field components E and B and frequency ω will cause a charged particle to oscillate with the electric field of the wave. For sufficiently intense waves, a relativistic calculation [15] shows that a charged particle will drift in the wave direction with a velocity v_P given by

$$v_P = \tfrac{1}{2}\left(\frac{\omega_{cs}}{\omega}\right)^2 \frac{\langle E \times B\rangle}{B^2} \quad \text{(SI)} \qquad (6.48)$$

$$= \tfrac{1}{2}\left(\frac{\omega_{cs}}{\omega}\right)^2 \frac{\langle E \times B\rangle}{cB^2} \quad \text{(cgs)} \qquad (6.49)$$

where $\langle \cdots \rangle$ denotes the average over one wave period.

Where the wave is spatially inhomogeneous, so that the electric field has a position dependent amplitude, then there is a net force f_p in the direction away from regions of high field intensity (even non-relativistically), given by [15, 70]

$$f_p = -\frac{1}{m_e}\left(\frac{q_s}{2\omega}\right)^2 \nabla E^2 \qquad (6.50)$$

where E denotes the electric field magnitude at the particle position. The force resulting from (6.50) is referred to as the ponderomotive force.

6.4.2 Adiabatic Invariants

6.4.2.1 Magnetic Moment
For a particle of charge q_s moving in a magnetic field which is non-uniform in time and space, the magnetic moment μ_{bs} is an adiabatic invariant of the motion:

$$\mu_{bs} = \frac{W_\perp}{B} = \text{constant} \qquad (6.51)$$

6.4.2.2 Longitudinal Invariant   The longitudinal invariant J_L is defined by

$$J_L = \oint v_{\parallel} ds = \int_a^b \left[\frac{2}{m}(W - \mu_m B) \right]^{1/2} ds \qquad (6.52)$$

where ds is an element of the guiding centre path, and the interval encompasses one complete cycle between reflection points a and b, that is, between zeros of the integrand.

6.4.3 Magnetic Mirror

A particle moving in a non-uniform, axisymmetric magnetic field may be reflected by the constancy of μ_{bs} (6.51), if the total energy of the particle $W = W_{\perp} + W_{\parallel}$ is a constant, and if the magnetic field increases to the point where $W_{\perp} = W$. Define the pitch angle θ by

$$\tan \theta = \frac{v_{\perp}}{v_{\parallel}} \qquad (6.53)$$

where $v_{\parallel}$ denotes the particle speed in the direction of the magnetic axis, and $v_{\perp}$ the speed in the orthogonal plane. A particle will be reflected if

$$\sin \theta > \left(\frac{B}{B_{max}} \right)^{1/2} \qquad (6.54)$$

where B_{max} is the maximum value of B. Such a configuration is termed a magnetic mirror. Two such configurations back-to-back form a magnetic bottle, in which particles are reflected from the strong magnetic fields at each end. If the minimum value of B between the mirrors is B_{min}, then the mirror ratio R is defined as

$$R = \frac{B_{max}}{B_{min}} \qquad (6.55)$$

and particles will be reflected if

$$\sin \theta \geq R^{-1/2} \qquad (6.56)$$

The propability p that particles will be lost from such a bottle, assuming a uniform distribution of particle velocities inside the bottle, is given by

$$p = 1 - \left(\frac{R-1}{R} \right)^{1/2} \qquad (6.57)$$

assuming no collective plasma effect. In practice, the plasma inside such a bottle relaxes collisionally in such a way that the component of the distribution depleted by mirror losses is continuously replaced, resulting eventually in the total loss of the plasma on a relaxation timescale.

6.5   TRANSPORT COEFFICIENTS

6.5.1   Fully Ionised Plasma, Zero Magnetic Field, Krook Operator

Using the Krook collision operator (5.30) for an unmagnetised plasma, and assuming a Maxwellian equilibrium, we have the following basic transport coefficients for particles of species s, having number density, charge and mass n_s, q_s and m_s respectively responding to the application of a uniform electric field:

$$\text{electrical conductivity: } \sigma_s = \frac{n_s q_s^2}{m_s \nu_c} \tag{6.58}$$

$$\text{particle mobility: } \mu_s = \frac{q_s}{m_s \nu_c} \tag{6.59}$$

$$\text{diffusion coefficient: } D_s = \frac{k_B T}{m_s \nu_c} \tag{6.60}$$

$$\text{thermal conductivity: } K_s = \frac{5 n_s k_B^2 T}{2 m_s \nu_c} \tag{6.61}$$

$$\text{viscosity: } \eta_{v,s} = \frac{n_s k_B T}{\nu_c} \tag{6.62}$$

Restrictions: Note that the collision frequency ν_c has been left unspecified in (6.58) -(6.62). The strong dependence of the collisional term on the particle speed means that ν_c can vary widely depending on the collision model. Note also that the plasma is assumed to be close to equilibrium (implicit in the construction of the Krook operator), and that the electric field is far below E_D, the threshold for electron runaway (see (9.112)).

6.5.2   Lorentzian and Spitzer Conductivity

6.5.2.1   Lorentz Conductivity   A Lorentzian plasma is one in which the ions form a stationary, infinitely heavy neutralising background charge distribution, and the mobile electrons interact solely with these ions; electron-electron interactions are ignored.

In this limit, an exact expression for the plasma conductivity, σ_L, for an unmagnetised plasma can be found by expanding a Fokker-Plank collision term for small departures from a Maxwellian equilibrium. The result is [86, 98]

$$\sigma_{\mathrm{L}} = \frac{2^{5/2}(4\pi\epsilon_0)^2 m_e^{1/2}(k_B T_e)^{3/2}}{Z e^2 (\pi m_e)^{3/2} \ln \Lambda} \quad \text{(SI)} \tag{6.63}$$

$$= \frac{2^{5/2} m_e^{1/2}(k_B T_e)^{3/2}}{Z e^2 (\pi m_e)^{3/2} \ln \Lambda} \quad \text{(cgs)} \tag{6.64}$$

where $\ln \Lambda$ is the Coulomb logarithm term (6.15), and Z is the charge on the ion, assuming only one species of ion.

6.5.2.2 Spitzer Conductivity The Lorentzian result (6.63) can be extended to accommodate electron-electron interactions, as well as ions of different species. The resulting Spitzer-Härm conductivity is given by [86, 98]

$$\sigma_{\mathrm{S}} = \gamma_E \sigma_{\mathrm{L}} \tag{6.65}$$

where γ_E is a correction factor dependent on the ionic charge; typical values were calculated by Spitzer and Härm:

Z	1	2	4	16	∞
γ_E	0.582	0.683	0.785	0.923	1.000

An empirical approximation to (6.65) is [98]

$$\sigma_{\mathrm{S}} = \left(0.295 + \frac{0.39}{0.85 + Z_{\mathrm{eff}}} \right)^{-1} \frac{n_e e^2}{m_e} \tau_e \tag{6.66}$$

where the collision time τ_e is defined by

$$\tau_e = 3(2\pi)^{3/2} \frac{\epsilon_0^2 m_e^{1/2} (k_B T_e)^{3/2}}{e^4 Z_{\mathrm{eff}} n_e \ln \Lambda} \quad \text{(SI)} \tag{6.67}$$

$$= \frac{3}{8} \left(\frac{2}{\pi} \right)^{1/2} \frac{m_e^{1/2} (k_B T_e)^{3/2}}{Z_{\mathrm{eff}} e^4 n_e \ln \Lambda} \quad \text{(cgs)} \tag{6.68}$$

and where the effective ion charge Z_{eff} is given by

$$Z_{\mathrm{eff}} = \frac{\sum_s n_s Z_s^2}{\sum_s n_s Z_s} \tag{6.69}$$

in which the sum is over all ionic species s with corresponding charge Z_s.

Restrictions: Note that in constructing (6.65) and (6.66) the dependence on Z of the Coulomb logarithm was ignored. Note also that it is assumed that the plasma is close to its Maxwellian equilibrium, and that the applied electric field is far below E_D (see (9.112)).

6.5.3 Fully Ionized and Magnetized Plasma: Braginskii Coefficients

A fully ionized plasma of two mutually interpenetrating fluids of ions and electrons, in the presence of a magnetic field, is analysed under the following assumptions [16, 17, 33, 98]:

- there is no neutral component; the plasma consists of ions, charge Ze, and electrons

- $m_i \gg m_e$ is exploited in the calculation of transport coefficients

- the equilibrium distribution function for ions and electrons is Maxwellian, though T_i is not necessarily equal to T_e

- the analysis is based on approximate solutions of the two-fluid Fokker-Planck equations, using the Rutherford scattering formula for collisions

- the minimum and maximum impact parameters used here are $\sim e^2/(mv^2)$ (6.11) and $\sim \lambda_d$ (6.10)

- the magnetic field does not influence the collision event itself, therefore the calculations are valid only for magnetic fields for which the Larmor radius is large compared to the Debye length

Braginskii's transport analysis yielded approximate forms for the electrical resistivity, thermoelectric and thermal conductivity tensors, each as a function of $\omega_c \tau$, where $\omega_c = \propto B$, and τ is a typical collision time.

The form of each transport quantity is presented as a rational function of $\omega_c \tau$ for ions and electrons, increasingly accurate for $\omega_c \tau \gg 1$, and valid over a range of Z. The numerical coefficients in the expansion are presented in Table 6.2; each numerical value is claimed to be better than 1% accurate [16], though the actual transport coefficients themselves can be in error by as much as 10-20% in the intermediate region $\omega_c \tau \sim 1$ [17].

The results are summarised in the following sections, using the notation

$$x_e = \omega_{ce} \tau_e \tag{6.70}$$

$$x_i = \omega_{ci} \tau_i \tag{6.71}$$

$$\Delta_e = x_e^4 + \delta_1 x_e^2 + \delta_0 \tag{6.72}$$

$$\Delta_i = x_i^4 + 2.70 x_i^2 + 0.677 \tag{6.73}$$

$$\tau_e = 3(2\pi)^{3/2} \epsilon_0^2 \frac{m_e^{1/2} (k_B T_e)^{3/2}}{e^4 Z^2 n_i \ln \Lambda} \quad \text{(SI)} \tag{6.74}$$

$$= \frac{3 m_e^{1/2} T_e^{3/2}}{4(2\pi)^{1/2} e^4 Z^2 n_i \ln \Lambda} \quad \text{(cgs)} \tag{6.75}$$

$$\tau_i = 3(2\pi)^{3/2} \epsilon_0^2 \frac{m_i^{1/2} (k_B T_i)^{3/2}}{e^4 Z^4 n_i \ln \Lambda} \quad \text{(SI)} \tag{6.76}$$

$$= \frac{3 m_i^{1/2} T_i^{3/2}}{4/(2\pi)^{1/2} e^4 Z^4 n_i \ln \Lambda} \quad \text{(cgs)} \tag{6.77}$$

and where subscripts $\parallel$, $\perp$ on vectors denote the directions parallel, and perpendicular, to that of the equilibrium magnetic field $\boldsymbol{B}_0$:

$$\boldsymbol{u}_{\parallel} = \boldsymbol{b}(\boldsymbol{b} \cdot \boldsymbol{u}) \tag{6.78}$$

$$\boldsymbol{u}_{\perp} = \boldsymbol{b} \times (\boldsymbol{b} \times \boldsymbol{u}) \tag{6.79}$$

$$\boldsymbol{b} = \boldsymbol{B}_0 / B_0 \tag{6.80}$$

Table 6.2: Braginskii numerical transport coefficients

	$Z = 1$	$Z = 2$	$Z = 3$	$z = 4$	$Z \to \infty$
α_0	0.5129	0.4408	0.3965	0.3752	0.2949
β_0	0.7110	0.9052	1.016	1.090	1.521
γ_0	3.162	4.890	6.064	6.920	12.47
δ_0	3.770	1.047	0.5814	0.4106	0.0961
δ_1	14.79	10.80	9.618	9.055	7.482
α_0'	1.837	0.5956	0.3515	0.2566	0.0678
α_0''	0.7796	0.3439	0.2400	0.1957	0.0940
α_1'	6.416	5.523	5.226	5.077	4.63
α_1''	1.704	1.704	1.704	1.704	1.704
β_0'	2.681	0.9473	0.5905	0.4478	0.1461
β_0''	3.053	1.784	1.442	1.285	0.877
β_1'	5.101	4.450	4.233	4.124	3.798
β_1''	3/2	3/2	3/2	3/2	3/2
γ_0'	11.92	5.118	3.525	2.841	1.20
γ_0''	21.67	15.37	13.53	12.65	10.23
γ_1'	4.664	3.957	3.721	3.604	3.25
γ_1''	5/2	5/2	5/2	5/2	5/2

6.5.3.1 Momentum Transfer From Ions To Electrons The rate of transfer of momentum from ions to electrons $\boldsymbol{R}$ consists of two contributions: a friction term $\boldsymbol{R}_u$ resulting from the relative velocity $\boldsymbol{u}$ between electrons and ions, and a term $\boldsymbol{R}_T$ pertaining to the presence of thermal gradients. The full result is

$$\boldsymbol{R} = \boldsymbol{R}_u + \boldsymbol{R}_T \tag{6.81}$$

$$\boldsymbol{R}_u = -\alpha_{\parallel}\boldsymbol{u}_{\parallel} - \alpha_{\perp}\boldsymbol{u}_{\perp} + \alpha_{\wedge}\boldsymbol{b} \times \boldsymbol{u} \tag{6.82}$$

$$\boldsymbol{R}_T = -\beta_{\parallel}^{uT}\nabla_{\parallel}T_e - \beta_{\perp}^{uT}\nabla_{\perp}T_e - \beta_{\wedge}^{uT}\boldsymbol{b} \times \nabla T_e \tag{6.83}$$

where

$$\alpha_{\parallel} = \frac{m_e n_e}{\tau_e}\alpha_0 \tag{6.84}$$

$$\alpha_{\perp} = \frac{m_e n_e}{\tau_e}\left(1 - \frac{\alpha_1' x_e^2 + \alpha_0'}{\Delta_e}\right) \tag{6.85}$$

$$\alpha_{\wedge} = \frac{m_e n_e}{\tau_e}\frac{x_e(\alpha_1'' x_e^2 + \alpha_0'')}{\Delta_e} \tag{6.86}$$

$$\beta_{\parallel}^{uT} = n_e\beta_0 \tag{6.87}$$

$$\beta_{\perp}^{uT} = n_e\frac{\beta_1' x_e^2 + \beta_0'}{\Delta_e} \tag{6.88}$$

$$\beta_{\wedge}^{uT} = n_e\frac{x_e(\beta_1'' x_e^2 + \beta_0'')}{\Delta_e} \tag{6.89}$$

6.5.3.2 Electron Heat Flux

The electron heat flux $\boldsymbol{q}_e$ also has a friction and thermal contribution:

$$\boldsymbol{q}_u^e = \boldsymbol{q}_{u,e} + \boldsymbol{q}_{T,e} \tag{6.90}$$

$$\boldsymbol{q}_{u,e} = \beta_{\parallel}^{Tu}\boldsymbol{u}_{\parallel} + \beta_{\perp}^{Tu}\boldsymbol{u}_{\perp} + \beta_{\wedge}^{Tu}\boldsymbol{b} \times \boldsymbol{u} \tag{6.91}$$

$$\boldsymbol{q}_{T,e} = -\kappa_{\parallel}^e\nabla_{\parallel}T_e - \kappa_{\perp}^e\nabla_{\perp}T_e - \kappa_{\wedge}^e\boldsymbol{b} \times \nabla T_e \tag{6.92}$$

where

$$\beta_{\zeta}^{Tu} = T_e\beta_{\zeta}^{uT}, \qquad \zeta = \parallel, \perp, \wedge \tag{6.93}$$

$$\kappa_{\parallel}^e = \frac{n_e T_e \tau_e}{m_e}\gamma_0 \tag{6.94}$$

$$\kappa_{\perp}^e = \frac{n_e T_e \tau_e}{m_e}\frac{\gamma_1' x_e^2 + \gamma_0'}{\Delta_e} \tag{6.95}$$

$$\kappa_{\wedge}^e = \frac{n_e T_e \tau_e}{m_e}\frac{x_e(\gamma_1'' x_e^2 + \gamma_0'')}{\Delta_e} \tag{6.96}$$

6.5.3.3 Ion Heat Flux The ion heat flux $\boldsymbol{q}_i$ arising from thermal gradients is the only significant transport process associated with the ions:

$$\boldsymbol{q}_i = -\kappa^i_\| \nabla_\| T_i - \kappa^i_\perp \nabla_\perp T_i + \kappa^i_\wedge \boldsymbol{b} \times \nabla T_i \tag{6.97}$$

where

$$\kappa^i_\| = 3.906 \frac{n_i T_i \tau_i}{m_i} \tag{6.98}$$

$$\kappa^i_\perp = \frac{n_i T_i \tau_i}{m_i} \frac{2x_i^2 + 2.645}{\Delta_i} \tag{6.99}$$

$$\kappa^i_\wedge = \frac{n_i T_i \tau_i}{m_i} \frac{x_i(5x_i^2/2 + 4.65)}{\Delta_i} \tag{6.100}$$

6.5.3.4 Resistivity The momentum transfer from ions to electrons as a result of relative velocity $\boldsymbol{u}$ between them (see Section 6.5.3.1) allows the definition of resistivity components:

$$\eta_\| = \eta_0 \alpha_0 \tag{6.101}$$

$$\eta_\perp = \eta_0 \left(1 - \frac{\alpha'_1 x_e^2 + \alpha'_0}{\Delta_e} \right) \tag{6.102}$$

$$\eta_\wedge = -\eta_0 \frac{x_e(\alpha''_1 x_e^2 + \alpha''_0)}{\Delta_e} \tag{6.103}$$

$$\eta_0 = \frac{m_e}{e^2 n_e \tau_e} \tag{6.104}$$

6.5.4 Corrections to Braginskii Coefficients

Recent calculations [33] show significant departures from Braginskii's standard transport coefficients. These newer calculations used a direct and accurate numerical simulation of the linearized Fokker-Planck equation for a fully ionised electron-ion plasma, for a continuum range of $\omega_c \tau$ (using least-squares curve fitting) and for a much larger range of Z. Numerical discrepancies of up to 65% in $\beta_\wedge$, $\kappa_\perp$ and $\kappa_\wedge$ are reported for $0.3 \lesssim \omega_c \tau \lesssim 30$.

Moreover, the asymptotic forms of certain transport coefficients were found to be different from that predicted by the Braginskii modelling:

$$\lim_{x_e \to \infty} \frac{\alpha_\wedge}{m_e n_e / \tau_e} \sim \begin{cases} x_e^{-1} & \text{from (6.85),} \\ x_e^{-2/3} & \text{from [33].} \end{cases} \tag{6.105}$$

$$\lim_{x_e \to \infty} \frac{\beta^{uT}_\perp}{n_e} \sim \begin{cases} x_e^{-2} & \text{from (6.88),} \\ x_e^{-5/3} & \text{from [33].} \end{cases} \tag{6.106}$$

6.5.5 Equal Mass Plasma Transport

The equivalent Braginskii transport analysis for an electron-positron (e-p) plasma is simpler due to the mass symmetry in the model [1], and the single collision time. The resistivity in the e-p plasma is significantly different:

$$\eta_{\parallel} = 0.5129\,\eta_0 \qquad \text{Braginskii, electron-ion result} \qquad (6.107)$$

$$\eta_{\parallel} = 0.1071\,\eta_0 \qquad \text{e-p equivalent} \qquad (6.108)$$

$$\eta_{\perp} = 0.5553\,\eta_0 \qquad \text{Braginskii, electron-ion result} \qquad (6.109)$$

$$\eta_{\perp} = 0.3845\,\eta_0 \qquad \text{e-p equivalent} \qquad (6.110)$$

$$\eta_{\wedge} = -0.1334\,\eta_0 \qquad \text{Braginskii, electron-ion result} \qquad (6.111)$$

$$\eta_{\wedge} = 0 \qquad \text{e-p equivalent} \qquad (6.112)$$

where the original Braginskii coefficients from Table 6.2 have been used for the case $Z = 1$, $\omega_{ce}\tau_e = 1$ to make a comparison with the computed e-p results of [1]. Further notable differences exist between electron-ion and electron-positron plasmas in the thermoelectric, thermal conductivity and diffusion tensors, across a wide range of $\omega_e\tau_e$; see [1] for extensive detail.

7

Plasma Waves

7.1 NOTATION

Symbol	Meaning	Ref
b_s	ratio of thermal to wave mode energy	(7.162)
c	speed of light in vacuo $= (\mu_0 \epsilon_0)^{-1/2}$	
c_a	Alfvén speed of single fluid plasma	(2.24)
c_{th}	sound speed in plasma gas	
$\mathbf{B}$	magnetic induction	
$\mathbf{E}$	electric field	
I_n	modified Bessel function of order n	
$\mathbf{J}$	current density	
$\mathbf{K}$	cold plasma dielectric tensor	(7.20)
m_s	mass of particle of species s	
n	refractive index of plasma, $= kc/\omega$	
n_c	cut-off density for an electron plasma	(7.56)
n_s	number density of particles of species s	
n_{tot}	total plasma density	(7.79)
p	scalar gas pressure of single-fluid plasma	(7.84)
p_s	scalar gas pressure of species s	
$\mathcal{P}$	total gas kinetic plus magnetic pressure	(7.128)
q_s	charge carried by particle of species s	
s	label defining species, e.g. ion (i) or electron (e)	
T	temperature of single-fluid plasma	
T_s	temperature of gas species s	
$\mathbf{u}$	bulk fluid plasma velocity	(7.81)
$\mathbf{u}_s$	velocity of species s	
$\mathcal{Z}$	plasma dispersion function	(10.7)
ϵ_0	vacuum permittivity	
$\boldsymbol{\epsilon}$	hot plasma dielectric tensor	(7.152)
η	fluid plasma resistivity	
μ_0	vacuum permeability	
ρ	mass density of single-fluid plasma	(7.80)
ρ_c	free charge density	
ρ_s	mass density of species s	
ω	frequency of electromagnetic wave	
ω_p	plasma frequency of the whole plasma	(2.6)
ω_{cs}	cylcotron frequency of species s	(2.7)
ω_{ps}	plasma frequency of species s	(2.6)

7.2   WAVES IN COLD PLASMAS

This section summarises small amplitude waves in a pressureless, magnetised plasma. The full model equations are given, followed by the dielectric tensor and general dispersion relation resulting from the plane wave solutions of the linearised equations. Special cases of local dispersion relations valid near particular frequencies are also given. There are a variety of excellent texts on these topics from which the results presented here are drawn [15, 19, 29, 50, 52, 88, 90]. The reader is encouraged to consult these texts for further details.

We shall assume a frequency hierarchy as follows [15, 88]:

$$\omega_{ci} \ll \omega_{ce} \leq \omega_{pe} \tag{7.1}$$

where the subscript i refers to any ion. Of course, in a plasma with several ion species, each ion cyclotron frequency varies with the ion mass.

7.2.1   Model Equations

The standard equations for a multi-component magnetised cold (pressureless) plasma are as follows (SI). For each particle species s, we have:

$$\frac{\partial n_s}{\partial t} + \nabla \cdot (n_s \boldsymbol{u}_s) = 0, \tag{7.2}$$

$$n_s m_s \frac{\partial \boldsymbol{u}_s}{\partial t} + n_s m_s (\boldsymbol{u}_s \cdot \nabla) \boldsymbol{u}_s = q_s (\boldsymbol{E} + \boldsymbol{u}_s \times \boldsymbol{B}), \tag{7.3}$$

$$\boldsymbol{J} = \sum_s n_s q_s \boldsymbol{u}_s. \tag{7.4}$$

These species equations, together with the Maxwell Equations:

$$\nabla \times \boldsymbol{E} = -\frac{\partial \boldsymbol{B}}{\partial t}, \tag{7.5}$$

$$\nabla \times \boldsymbol{B} = \mu_0 \boldsymbol{J} + \mu_0 \epsilon_0 \frac{\partial \boldsymbol{E}}{\partial t}, \tag{7.6}$$

$$\nabla \cdot \boldsymbol{E} = \rho_c / \epsilon_0, \tag{7.7}$$

$$\nabla \cdot \boldsymbol{B} = 0, \tag{7.8}$$

provide the mathematical framework for describing the dynamics of cold plasmas.

These equations may be cast in cgs-Gaussian units as follows:

$$\frac{\partial n_s}{\partial t} + \nabla \cdot (n_s \boldsymbol{u}_s) = 0, \tag{7.9}$$

$$n_s m_s \frac{\partial \boldsymbol{u}_s}{\partial t} + n_s m_s (\boldsymbol{u}_s \cdot \nabla) \boldsymbol{u}_s = q_s (\boldsymbol{E} + \boldsymbol{u}_s \times \boldsymbol{B}/c), \qquad (7.10)$$

with the appropriate Maxwell equations:

$$c\nabla \times \boldsymbol{E} = -\frac{\partial \boldsymbol{B}}{\partial t}, \qquad (7.11)$$

$$c\nabla \times \boldsymbol{B} = 4\pi \boldsymbol{J} + \frac{\partial \boldsymbol{E}}{\partial t}, \qquad (7.12)$$

$$\nabla \cdot \boldsymbol{E} = 4\pi/\rho_c, \qquad (7.13)$$

$$\nabla \cdot \boldsymbol{B} = 0. \qquad (7.14)$$

7.2.2   Cold Plasma Variable Dependencies

For small amplitude disturbances varying as $\exp[i(\boldsymbol{k} \cdot \boldsymbol{r} - \omega t)]$ about a static equilibrium, and assuming a stationary ion background for simplicity, the various relationships between the (electron) plasma variables can be written as follows:

$$\frac{n}{n_0} = \frac{\boldsymbol{k} \cdot \boldsymbol{u}}{\omega} \qquad (7.15)$$

$$\boldsymbol{E} = \mathbf{M}\boldsymbol{u} \qquad (7.16)$$

$$\boldsymbol{B} = \lambda(\boldsymbol{k} \times \boldsymbol{u}) \qquad (7.17)$$

where subscript 0 on a variable denotes an equilibrium quantity, $\boldsymbol{k}$ is confined to the x-z plane, $\boldsymbol{B}_0 = \hat{\boldsymbol{z}} B_0$, $\mathbf{M}$ is the matrix given by

$$\mathbf{M} = \begin{bmatrix} -i\omega & \omega_{ce} & 0 \\ -\omega_{ce} & -i\omega & 0 \\ 0 & 0 & -i\omega \end{bmatrix}, \qquad (7.18)$$

and

$$\lambda = i\mu_0 n_0 e \left(k^2 + \omega^2/c^2\right)^{-1}. \qquad (7.19)$$

7.2.3   Dielectric Tensor for a Cold Magnetised Plasma

Small amplitude disturbances of the model equations, assuming that the uniform magnetic field points in the z-direction, yield the following expression for the dielectric tensor $\mathbf{K}$, defined by $\boldsymbol{J} = \mathbf{K}\boldsymbol{E}$:

$$\mathbf{K} = \begin{bmatrix} S & -iD & 0 \\ iD & S & 0 \\ 0 & 0 & P \end{bmatrix} \qquad (7.20)$$

where

$$S = (R + L)/2 \tag{7.21}$$

$$D = (R - L)/2 \tag{7.22}$$

$$R = 1 - \sum_s \frac{\omega_{ps}^2}{\omega^2} \frac{\omega}{\omega + \omega_{cs}} \tag{7.23}$$

$$L = 1 - \sum_s \frac{\omega_{ps}^2}{\omega^2} \frac{\omega}{\omega - \omega_{cs}} \tag{7.24}$$

$$P = 1 - \sum_s \frac{\omega_{ps}^2}{\omega^2} = 1 - \frac{\omega_p^2}{\omega^2} \tag{7.25}$$

For the simple case of an electron-proton plasma, (7.23) and (7.24) can be written as

$$R = 1 - \frac{\omega_p^2}{(\omega + \omega_{ci})(\omega + \omega_{ce})} \tag{7.26}$$

$$L = 1 - \frac{\omega_p^2}{(\omega - \omega_{ci})(\omega - \omega_{ce})} \tag{7.27}$$

Note that the cyclotron frequency ω_{cs} of species s carries the sign of the charge on species s. The general solution for linear waves in a cold magnetised plasma can be expressed entirely in terms of a matrix equation involving only the perturbed electric field:

$$\begin{bmatrix} S - n^2 \cos^2 \theta & -iD & n^2 \cos \theta \sin \theta \\ iD & S - n^2 & 0 \\ n^2 \cos \theta \sin \theta & 0 & P - n^2 \sin^2 \theta \end{bmatrix} \begin{bmatrix} E_x \\ E_y \\ E_z \end{bmatrix} = 0, \tag{7.28}$$

where $n = kc/\omega$ is the refractive index.

7.2.4 General Dispersion Relation

The general dispersion relation for waves in a uniform, magnetised cold plasma propagating at angle θ to the direction of the magnetic field is derived from the determinant of the matrix in (7.28):

$$(S \sin^2 \theta + P \cos^2 \theta)n^4 - (RL \sin^2 \theta + PS(1 + \cos^2 \theta))n^2 + PRL = 0 \tag{7.29}$$

An alternative form is

$$\tan^2 \theta = -\frac{P(n^2 - R)(n^2 - L)}{(Sn^2 - RL)(n^2 - P)}. \tag{7.30}$$

The electric field components transverse to the equilibrium magnetic field are related by

$$\frac{iE_x}{E_y} = \frac{n^2 - S}{D}. \tag{7.31}$$

There is also a simple relationship between the transverse velocity components:

$$i\frac{v_{xs}}{v_{ys}} = \frac{iE_x/E_y - \omega_{cs}/\omega}{1 - (\omega_{cs}/\omega)(iE_x/E_y)}. \tag{7.32}$$

7.2.4.1 Parallel Propagation

Plasma Oscillations   A solution to the dispersion relation for $\theta = 0$ is

$$P = 0 \tag{7.33}$$
$$\omega^2 = \omega_p^2 \tag{7.34}$$

which corresponds to electrostatic plasma oscillations. Here the electric field is aligned with the magnetic field,

$$E_x = E_y = 0 \tag{7.35}$$

the oscillation is purely longitudinal,

$$v_{xs} = v_{ys} = 0 \tag{7.36}$$

$$v_{zs} = -\frac{q_s}{i\omega_{ps}m_s}E_z \tag{7.37}$$

and at the plasma frequency,

$$\omega = \omega_p. \tag{7.38}$$

There is a maximum amplitude for such oscillations, revealed in a treatment of the non-linear cold non-relativistic electron plasma oscillation [28, 68]:

$$E_{max} = \frac{n_0 e}{k\epsilon_0} \quad \text{(SI)} \tag{7.39}$$

$$E_{max} = \frac{n_0 e}{4\pi k} \quad \text{(cgs)} \tag{7.40}$$

where k is the wave-number corresponding to a longitudinal electric field disturbance of the cold electron fluid of the form

$$E_z = E_0 \sin(kx_0 - \omega_p t) \tag{7.41}$$

with x_0 denoting the initial position of a fluid element, about which such oscillations are taking place. The condition (7.39) corresponds to the breakdown of the harmonic behaviour.

For relativistic cold electron plasma oscillations, (7.39) can be generalised to

$$E_{max} = \sqrt{2}\,\frac{m_e c \omega_{pe}}{e}\,(\gamma - 1)^{1/2}\,, \tag{7.42}$$

$$\gamma = \left(1 - \frac{\omega_p}{kc}\right)^{1/2} \tag{7.43}$$

In equal-mass plasmas, such as electron-positron plasmas [27], the electrostatic oscillation is not a stable mode, since there is no static ion background. The instability develops in the number densities of each species, causing large density gradients to evolve which deplete the total plasma density in the central regions of the oscillation and lead to a variation in the local plasma frequency across the oscillation region, exacerbating the growth of density spikes.

Circularly polarized waves For waves propagating parallel to the magnetic field ($\theta = 0$), the solutions to the dispersion relation are

$$n^2 = R \quad \text{(RCP)} \tag{7.44}$$

$$n^2 = L \quad \text{(LCP)}, \tag{7.45}$$

which correspond to circularly polarized waves. The polarization is determined from the usual radio convention:

$$\frac{iE_x}{E_y} = +1 \quad \text{(Right Circular polarization)} \tag{7.46}$$

$$\frac{iE_x}{E_y} = -1 \quad \text{(Left Circular polarization)} \tag{7.47}$$

The general solutions (7.44) and (7.45) have particular forms in certain frequency ranges.

Faraday Rotation The circularly polarized modes described by (7.44) and (7.45) have different phase speeds, due to their different refractive indices. This effect can be exploited as a plasma diagnostic, since any plane polarized wave propagating parallel to the magnetic field can be expressed as a superposition of the circularly polarized modes. Consequently the direction of the plane of polarization must vary as the wave propagates, since each circular component travels at a different phase speed. In general terms this gives rise to a rotation of the plane of polarization, with a rotation angle ψ given by [46]:

$$\psi = \frac{1}{2\omega^2 c} \int_0^L \omega_p^2 \omega_{ce} \cos\theta \, \mathrm{d}z \tag{7.48}$$

where it is assumed that the plasma is a cold electron plasma, $\omega \gg \omega_{ce}$ is the wave (circular) frequency, θ is the angle between the magnetic field and propagation directions (assumed small), and the plasma extends a distance L along the propagation direction.

The utility of Faraday rotation in an astrophysical context lies not in calculating the absolute rotation angle, since the initial direction of the plane of rotation is not known, but rather in exploiting (7.48) as a function of frequency (or wavelength). Defining the *rotation measure RM* by [46]:

$$RM = \psi \nu^2 \tag{7.49}$$

$$= \frac{1}{8\pi^2 c} \int_0^L \frac{\omega_p^2 \omega_{ce} \cos\theta}{(1 - \omega_p^2/\omega^2)^{1/2}} dz \tag{7.50}$$

so that $\psi = RM/\nu^2$ where nu is the wave frequency. A measure of the change in ψ at two frequencies then yields RM, and consequently information about the line-of-sight integrated product of density and magnetic field.

A related diagnostic is the dispersion measure, which yields the line-of-sight electron number density, and is defined as follows. Assuming $\omega \gg \omega_p, \omega_{ce}$, the dispersion relation can be approximated by

$$k \approx \frac{\omega}{c} \left(1 - \frac{\omega_p^2}{2\omega^2} \right) \tag{7.51}$$

where k is the wavenumber. Hence the total propagation time t_{prop} of a signal traversing a length L of plasma can be approximated by

$$t_{\text{prop}} \approx \frac{L}{c} + \frac{1}{2c\omega^2} \int_0^L \omega_p^2 dz \tag{7.52}$$

$$= \frac{d}{c} + \frac{DM}{\nu^2} \tag{7.53}$$

where the *dispersion measure DM* is defined by

$$DM = \frac{1}{8\pi^2 c} \int_0^L \omega_p^2 dz \tag{7.54}$$

The ratio of (7.50) and (7.54) is sometimes used to infer a mean density-weighted magnetic field.

7.2.4.2 Resonances and Cut-Offs: Parallel Propagation

The following definitions are useful in discussing cold plasma waves.

A *cut-off* frequency is one for which the wave-number is zero.

A *resonance* frequency is one for which the group velocity of a wave is zero.

For an unmagnetised plasma, waves cannot propagate unless

$$\omega \geq \omega_p \tag{7.55}$$

and so ω_p is the cut-off frequency for an unmagnetised plasma. Associated with this is the concept of the cut-off density, n_c, defined for an electron plasma as

$$n_c = \frac{\omega^2 \epsilon_0 m_e}{e^2} \quad \text{(SI)} \tag{7.56}$$

$$= \frac{4\pi \omega^2 m_e}{e^2} \quad \text{(cgs)} \tag{7.57}$$

At the cyclotron frequencies $\omega = \omega_{cs}$ there are wave resonances. For an ion-electron plasma, the LCP solution ($n^2 = L$) has a resonance at $\omega = \omega_{ci}$, with $\omega = \omega_{ce}$ being a resonance for the RCP mode ($n^2 = R$). A proper treatment of the plasma response at these frequencies requires a kinetic theory approach.

Note also that the right and left circularly polarized modes also have cut-off frequencies, i.e. those frequencies which define where the wave-number is zero.

Together, the resonances and cut-offs define different band-gaps for each polarization where no wave solution exists.

There is no LCP wave for frequencies in the range

$$\omega_{ci} \leq \omega \leq \omega_{CL} \tag{7.58}$$

and there is no RCP wave for frequencies in the range

$$\omega_{ce} \leq \omega \leq \omega_{CR} \tag{7.59}$$

where the cut-off frequencies ω_{CL}, ω_{CR} for the LCP and RCP waves respectively are given as:

$$\omega_{CL} \approx [(\omega_{ce}^2 + 4\omega_p^2)^{1/2} - \omega_{ce}]/2 \tag{7.60}$$

$$\omega_{CR} \approx [(\omega_{ce}^2 + 4\omega_p^2)^{1/2} + \omega_{ce}]/2. \tag{7.61}$$

Note that we have assumed $\omega_{ce} < \omega_p$. At high frequencies, the LCP and RCP plasma modes resemble circularly polarized electromagnetic vacuum waves.

Alfvén Wave For ultra low frequency waves, such that $\omega \ll \omega_{ci}$, we can simplify the expressions (7.23), (7.24) as follows:

$$R \approx L \approx 1 + c^2/c_a^2 \tag{7.62}$$

leading to the dispersion relation for Alfvén waves:

$$\omega^2 \approx \frac{k^2 c^2}{1 + c^2/c_a^2} \approx k^2 c_a^2. \tag{7.63}$$

Ion Cyclotron Wave   As $\omega \to \omega_{ci}$ from below, the dispersion relation changes significantly from (7.63):

$$\omega^2 \approx \frac{k^2 c^2}{2\omega_{pi}^2} \left(\omega_{ci}^2 - \omega^2 \right). \tag{7.64}$$

This electromagnetic mode is termed the ion cyclotron wave, and is in general elliptically polarized. The group velocity of this wave tends to zero at the ion cyclotron frequency, and so is resonant at ω_{ci}. Note that (7.64) can be generalised for propagation at an arbitrary angle θ:

$$\omega^2 \approx \frac{k^2 c^2}{\omega_{pi}^2} \left(\omega_{ci}^2 - \omega^2 \right) \frac{\cos^2 \theta}{1 + \cos^2 \theta} \tag{7.65}$$

whistler waves   At intermediate frequencies, namely $\omega_i < \omega < \omega_e, \omega_p$, the local approximation is the whistler mode:

$$\omega^2 \approx \frac{k^2 c^2}{1 + \omega_p^2/(\omega \omega_{ce})}. \tag{7.66}$$

Assuming that $\omega_p \leq \omega_{ce}$ only the RCP solution exists, and so whistlers are right circularly polarized modes.

Electron Cyclotron Wave   The RCP mode has a resonance at the electron cyclotron frequency, which can be generalised to the following expression for propagation at an angle θ to the equilibrium magnetic field direction:

$$n^2 \approx 1 - \frac{\omega_{pe}^2}{\omega(\omega + \omega_{ce} \cos \theta)}. \tag{7.67}$$

This mode is referred to as the electron cyclotron mode.

7.2.4.3   Perpendicular Propagation

For waves propagating perpendicular to the equilibrium magnetic field direction, that is for $\theta = \pi/2$, the dispersion relation has the general solutions

$$n^2 = P \qquad \text{Ordinary mode} \tag{7.68}$$

$$n^2 = \frac{RL}{S} \qquad \text{Extraordinary mode} \tag{7.69}$$

The Ordinary mode (O-Mode) is purely transverse and linearly polarized, as can be deduced from the third row of (7.28); it does not depend on the magnetic field. The physical reason for this is that the electric field for this mode lies in the same direction as the magnetic field and so particles are accelerated parallel to $\boldsymbol{B_0}$.

The extraordinary mode (X-Mode) has an electric field which lies in the x-y plane, producing a mode which is elliptically polarized in that plane:

$$iE_x/E_y = -D/S \tag{7.70}$$

7.2.4.4 Resonances and Cut-Offs: Perpendicular Propagation   The only resonance for perpendicular propagation occurs for the X-mode when $S = 0$, giving rise to the hybrid resonances (using the frequency hierarchy of (7.1)):

$$\omega_u^2 \approx \omega_p^2 + \omega_{ce}^2 \quad \text{Upper Hybrid} \qquad (7.71)$$

$$\omega_l^2 \approx \frac{\omega_p^2 \omega_{ce} \omega_{ci}}{\omega_p^2 + \omega_{ce}^2} \quad \text{Lower Hybrid} \qquad (7.72)$$

Note that from (7.70), as $\omega \to \omega_{u,l}, iE_x/E_y \to \infty$, showing that the X-mode becomes predominantly longitudinal near the hybrid resonances.

The X-mode shares the same cut-offs as the circularly polarized modes, namely $R = 0$ and $L = 0$. The O-mode however cannot propagate below ω_p.

7.2.4.5 Fast Alfvén Wave   For ultra-low frequency waves $\omega \ll \omega_{ci}$ propagating at $\theta = \pi/2$, only the X-mode is possible, and the dispersion relation becomes once more (cf (7.63)):

$$\omega^2 \approx \frac{k^2 c^2}{1 + c^2/c_a^2} \approx k^2 c_a^2. \qquad (7.73)$$

However, there is a very significant difference here from the parallel case, in that from (7.70), since $R \approx L$, E_y dominates in this Alfvén wave, making it electromagnetically transverse, but mechanically longitudinal, since from (7.16) $v_x \gg v_y$. This means that the perturbed magnetic field is parallel or anti-parallel to the equilibrium magnetic field, making the wave magnetically compressional.

7.2.5 Equal-Mass Cold Plasmas

For the special case where the positively and negatively charged species have the same mass, for example in an electron-positron plasma, there are simplifications arising directly from the symmetry. In particular, the dielectric tensor **K** (7.20) has no off-diagonal components [87], since $D = 0$. Consequently there is no Faraday rotation, and no 'whistler' type wave for parallel propagation.

7.3 FLUID PLASMAS

The plasma here is treated as a continuum of charged matter density with significant pressure effects. Separate equations for the fluid properties of the ions and electrons can be derived from moments of the appropriate kinetic equations; appropriate details can be found in advanced textbooks such as [15, 21, 23, 44, 98].

7.3.1   Hydromagnetic Equations

Here we assume, for simplicity, a two-component electron-ion fully ionised and magnetised plasma. The continuum equations for each fluid species are obtained via moments of the kinetic equation (5.8) (see (5.1) - (5.5)).

zeroth moment

$$\frac{\partial n_s}{\partial t} + \nabla \cdot (n_s \boldsymbol{u_s}) = \left(\frac{\partial n}{\partial t}\right)_c \tag{7.74}$$

where the term on the right denotes the effect of collisions on the number density of each species.

first moment   The momentum equation for each fluid species s is

$$\rho_s \left(\frac{\partial \boldsymbol{u_s}}{\partial t} + (\boldsymbol{u_s} \cdot \nabla)\boldsymbol{u_s}\right) = q_s n_s \left(\boldsymbol{E} + \boldsymbol{u_s} \times \boldsymbol{B}\right) - \nabla p_s + \rho_s \boldsymbol{F}_{ext} + \boldsymbol{P}_{ss'} \qquad \text{(SI)}$$
$$\tag{7.75}$$

$$= q_s n_s \left(\boldsymbol{E} + \frac{\boldsymbol{u_s} \times \boldsymbol{B}}{c}\right) - \nabla p_s + \rho_s \boldsymbol{F}_{ext} + \boldsymbol{P}_{ss'} \quad \text{(cgs)}$$
$$\tag{7.76}$$

where $\boldsymbol{F}_{ext}$ denotes an external force term, and $\boldsymbol{P}_{ss'}$ denotes the momentum transfer between the gas species. Note that (7.75) assumes a scalar pressure; the most general form incorporates a pressure tensor $\mathbf{P}_s$, with ∇p_s being replaced by $\nabla \cdot \mathbf{P}_s$.

The momentum equations for each species may be combined to derive a generalised Ohm's law:

$$\boldsymbol{E} + \boldsymbol{u} \times \boldsymbol{B} = \eta \boldsymbol{J} + \frac{1}{e\rho}\left[\frac{m_i m_e}{e}\frac{\partial \boldsymbol{J}}{\partial t}\right]$$
$$+ \frac{1}{e\rho}\left[m_e \nabla p_i - m_i \nabla p_e + (m_i - m_e)\boldsymbol{J} \times \boldsymbol{B}\right] \qquad \text{(SI)} \quad (7.77)$$

$$\boldsymbol{E} + \frac{\boldsymbol{u} \times \boldsymbol{B}}{c} = \eta \boldsymbol{J} + \frac{1}{e\rho}\left[\frac{m_i m_e}{e}\frac{\partial \boldsymbol{J}}{\partial t}\right]$$
$$+ \frac{1}{e\rho}\left[m_e \nabla p_i - m_i \nabla p_e + \left(\frac{m_i - m_e}{c}\right)\boldsymbol{J} \times \boldsymbol{B}\right] \quad \text{(cgs)}$$
$$\tag{7.78}$$

Restrictions

- plasma is fully ionised, and overall neutral;
- each fluid has a scalar pressure;

- viscosity is ignored;

- momentum exchange between ions (protons here) and electrons is assumed proportional to their relative mean velocities.

7.3.2 Single Fluid MHD Plasma

The main results for small amplitude waves in single-fluid magnetohydrodynamics (MHD) are summarised here. The model equations are stated, together with validity criteria, and the general dispersion relation is presented and the wave properties discussed.

single fluid variables   The single fluid macroscopic variables for an electron-ion plasma are defined as follows:

$$n_{tot} = n_i + n_e \qquad\qquad \text{number density} \qquad (7.79)$$

$$\rho = \rho_i + \rho_e = n_i m_i + n_e m_e \qquad \text{mass density} \qquad (7.80)$$

$$\boldsymbol{u} = \left(\rho_i \boldsymbol{u}_i + \rho_e \boldsymbol{u}_e\right)/\rho \qquad \text{bulk velocity} \qquad (7.81)$$

$$q = q_i n_i - e n_e \qquad\qquad \text{charge density} \qquad (7.82)$$

$$\boldsymbol{J} = n_i q_i \boldsymbol{u}_i - n_e e \boldsymbol{u}_e \qquad \text{total current} \qquad (7.83)$$

$$p = p_i + p_e \qquad\qquad\qquad \text{total pressure} \qquad (7.84)$$

The standard equations for a single fluid plasma are as follows (in SI):

$$\frac{\partial \rho}{\partial t} + \nabla \cdot (\rho \boldsymbol{u}) = 0 \qquad (7.85)$$

$$\rho \left(\frac{\partial}{\partial t} + \boldsymbol{u} \cdot \nabla \right) \boldsymbol{u} = -\nabla p + \boldsymbol{J} \times \boldsymbol{B} \qquad (7.86)$$

$$\left(\frac{\partial}{\partial t} + \boldsymbol{u} \cdot \nabla \right) \left(p \rho^{-5/3} \right) = \frac{2}{3} \rho^{-5/3} \eta J^2 \qquad (7.87)$$

$$\boldsymbol{E} + \boldsymbol{u} \times \boldsymbol{B} = \eta \boldsymbol{J} \ \text{(Ohm's Law)} \qquad (7.88)$$

together with the (reduced) electromagnetic equations (again in SI):

$$\nabla \times \boldsymbol{B} = \mu_0 \boldsymbol{J}, \qquad (7.89)$$

$$\nabla \times \boldsymbol{E} = -\frac{\partial \boldsymbol{B}}{\partial t}, \qquad (7.90)$$

$$\nabla \cdot \boldsymbol{B} = 0. \qquad (7.91)$$

In cgs units, the respective equations are:

$$\rho\left(\frac{\partial}{\partial t} + \boldsymbol{u}\cdot\nabla\right)\boldsymbol{u} = -\nabla p + \boldsymbol{J}\times\boldsymbol{B}/c, \qquad (7.92)$$

$$\boldsymbol{E} + \boldsymbol{u}\times\boldsymbol{B}/c = \eta\boldsymbol{J}, \qquad (7.93)$$

$$\nabla\times\boldsymbol{B} = 4\pi\boldsymbol{J}/c, \qquad (7.94)$$

$$\nabla\times\boldsymbol{E} = -\frac{1}{c}\frac{\partial\boldsymbol{B}}{\partial t}, \qquad (7.95)$$

$$\nabla\cdot\boldsymbol{B} = 0. \qquad (7.96)$$

Note that the density equation (7.85) and the energy (entropy) equation (7.87) are the same in both sets of units.

The macroscopic single fluid variables are defined in terms of the plasma components in the following equations, where we have assumed an electron-proton plasma for simplicity:

$$n_{tot} = n_i + n_e \qquad (7.97)$$

$$\rho = m_i n_i + m_e n_e \qquad (7.98)$$

$$\boldsymbol{J} = e\left(n_i\boldsymbol{u}_i - n_e\boldsymbol{u}_e\right) \qquad (7.99)$$

$$p = 3Nk_BT \qquad (7.100)$$

It is vital to realise that the derivation of (7.85) to (7.88) is dependent upon many simplifying assumptions, the details of which are not recorded here, but the consequences of which can be encapsulated as follows [15]:

$$\frac{u}{c} \ll 1 \qquad \text{neglect displacement current} \qquad (7.101)$$

$$\frac{\omega}{\omega_p} \ll \frac{u}{c} \qquad \text{neglect } \partial_t\boldsymbol{J} \text{ in Ohm's Law} \qquad (7.102)$$

$$\frac{\omega\omega_{ce}}{\omega_p^2} \ll \left(\frac{u}{c}\right)^2 \qquad \text{neglect Hall term in Ohm's Law} \qquad (7.103)$$

$$\frac{\omega}{\omega_i} \ll \left(\frac{u}{c_{th}}\right)^2 \qquad \text{neglect pressure gradient in Ohm's Law} \qquad (7.104)$$

In general terms, the MHD description of a plasma is suitable only for non-relativistic, low frequency, long-wavelength perturbations which do not give rise to electromagnetic effects. Despite these restrictions, the model enjoys widespread popularity. The special case of $\eta \to 0$ is referred to as ideal MHD, in which the plasma is perfectly conducting.

7.3.3   Variable Dependencies in Ideal MHD

Assume a homogeneous, static and stationary equilibrium, and take all linear perturbations to vary as $\exp[\mathrm{i}(\boldsymbol{k}\cdot\boldsymbol{r} - \omega t)]$. Then there are several useful inter-relations between perturbed variables (SI units):

$$(\boldsymbol{k} \cdot \boldsymbol{u}) = \frac{\omega}{\rho_0} \rho \tag{7.105}$$

$$\omega \rho_0 \boldsymbol{u} = \boldsymbol{k} \left(p + \boldsymbol{B}_0 \cdot \boldsymbol{B}/\mu_0 \right) - (\boldsymbol{k} \cdot \boldsymbol{B}_0) \, \boldsymbol{B}/\mu_0 \tag{7.106}$$

$$\omega \boldsymbol{B} = - (\boldsymbol{k} \cdot \boldsymbol{B}_0) \, \boldsymbol{u} + (\boldsymbol{k} \cdot \boldsymbol{u}) \, \boldsymbol{B}_0 \tag{7.107}$$

where subscript 0 denotes an equilibrium quantity. In cgs units, only (7.106) is different:

$$\omega \rho_0 \boldsymbol{u} = \boldsymbol{k} \left(p + \boldsymbol{B}_0 \cdot \boldsymbol{B}/(4\pi) \right) - (\boldsymbol{k} \cdot \boldsymbol{B}_0) \, \boldsymbol{B}/(4\pi) \tag{7.108}$$

Equations (7.105-7.107) can be manipulated to yield other useful general relations:

$$\boldsymbol{k} \cdot \boldsymbol{B} = 0 \tag{7.109}$$

$$\rho = \frac{k^2}{\omega^2} \mathcal{P} \tag{7.110}$$

$$\boldsymbol{u} \cdot \boldsymbol{B}_0 = \frac{\boldsymbol{k} \cdot \boldsymbol{B}_0}{\omega \rho_0} p \tag{7.111}$$

$$\boldsymbol{u} \cdot \boldsymbol{B} = -\frac{k c_a^2 \cos \theta}{B_0 \omega} B^2 \tag{7.112}$$

$$\boldsymbol{u} \times \boldsymbol{B} = \frac{\mathcal{P}}{\rho_0} \frac{(\boldsymbol{k} \cdot \boldsymbol{u})(\boldsymbol{k} \times \boldsymbol{B}_0)}{\omega^2 - k^2 c_a^2 \cos^2 \theta} \tag{7.113}$$

where θ is the angle between $\boldsymbol{k}$ and $\boldsymbol{B}_0$, and where the total plasma pressure perturbation $\mathcal{P}$ is given by

$$\mathcal{P} = p + \boldsymbol{B}_0 \cdot \boldsymbol{B}/\mu_0 \qquad \text{linear perturbation (SI)} \tag{7.114}$$

$$\mathcal{P} = p + \boldsymbol{B}_0 \cdot \boldsymbol{B}/(4\pi) \qquad \text{linear perturbation (cgs)} \tag{7.115}$$

$$\tag{7.116}$$

7.3.4   General Dispersion Relation: Ideal MHD

The general dispersion relation for waves in a uniform ideal MHD plasma, propagating at an angle θ to the equilibrium magnetic field, is given by

$$\left(\omega^2 - k^2 c_a^2 \cos^2 \theta \right) \left(\omega^4 - k^2 \left(c_{th}^2 + c_a^2 \right) \omega^2 + k^4 c_{th}^2 c_a^2 \cos^2 \theta \right) = 0. \tag{7.117}$$

The first factor in (7.117) defines the Alfvén mode; the second factor, the magnetosonic modes.

7.3.4.1 Alfvén Wave   The Alfvén mode in MHD is a transverse, incompressible mhd wave, with the following properties:

$$\omega = \pm kc_a \cos\theta \qquad \text{dispersion relation} \qquad (7.118)$$

$$\boldsymbol{k} \cdot \boldsymbol{u} = 0 \qquad \text{incompressible} \qquad (7.119)$$

$$\boldsymbol{k} \cdot \boldsymbol{B} = 0 \qquad \text{transverse magnetic} \qquad (7.120)$$

$$\boldsymbol{E} = 0 \qquad \text{no electric field} \qquad (7.121)$$

$$p = -\boldsymbol{B} \cdot \boldsymbol{B}_0/\mu_0 \qquad \text{(SI)} \qquad (7.122)$$

$$p = -\boldsymbol{B} \cdot \boldsymbol{B}_0/(4\pi) \qquad \text{(cgs)} \qquad (7.123)$$

$$\mathcal{P} = 0 \qquad \text{constant total pressure} \qquad (7.124)$$

$$\boldsymbol{u} = \pm(\mu_0\rho_0)^{-1/2}\boldsymbol{B} \qquad \boldsymbol{u} \text{ and } \boldsymbol{B} \text{ aligned} \qquad (7.125)$$

$$\frac{1}{2}\rho_0 u^2 = \frac{1}{2}B^2/\mu_0 \qquad \text{equipartition of energy density (SI)} \qquad (7.126)$$

$$\frac{1}{2}\rho_0 u^2 = \frac{1}{2}B^2/(4\pi) \qquad \text{equipartition of energy density (cgs)} \qquad (7.127)$$

Note that (7.118-7.127) are valid for finite amplitude MHD Alfvén waves, with the total pressure statement restricting the maximum amplitude of the Alfvén wave:

$$\mathcal{P} = p_{\text{total}} + B^2_{\text{total}}/(2\mu_0) = \text{constant} \qquad \text{(SI)} \qquad (7.128)$$

$$\mathcal{P} = p_{\text{total}} + B^2_{\text{total}}/(8\pi) = \text{constant} \qquad \text{(cgs)} \qquad (7.129)$$

7.3.4.2 Magnetosonic Modes   The magnetosonic (or magneto-acoustic) modes are waves satisfying the dispersion relations

$$\left(\frac{\omega}{k}\right)^2 = \frac{1}{2}(c_{th}^2 + c_a^2) \pm \frac{1}{2}\left[(c_{th}^2 + c_a^2)^2 - 4c_{th}^2 c_a^2 \cos^2\theta\right]^{1/2} \qquad (7.130)$$

where the + solution denotes the fast magnetosonic mode, and −, the slow magnetosonic mode.

The relation between the magnetic and kinetic pressure contributions is as follows:

$$\frac{\boldsymbol{B} \cdot \boldsymbol{B}_0}{\mu_0} = \frac{c_a^2}{c_{th}^2}\left(1 - \frac{k^2 c_{th}^2}{\omega^2}\cos^2\theta\right)p \qquad \text{(SI)} \qquad (7.131)$$

$$\frac{\boldsymbol{B} \cdot \boldsymbol{B}_0}{4\pi} = \frac{c_a^2}{c_{th}^2}\left(1 - \frac{k^2 c_{th}^2}{\omega^2}\cos^2\theta\right)p \qquad \text{(cgs)} \qquad (7.132)$$

This means that

$$\text{fast magnetosonic} \quad \omega^2 > k^2 c_{th}^2 \cos^2 \theta \quad \text{pressures reinforce} \quad (7.133)$$

$$\text{slow magnetosonic} \quad \omega^2 < k^2 c_{th}^2 \cos^2 \theta \quad \text{pressures oppose} \quad (7.134)$$

7.4 WAVES IN HOT PLASMAS

This section summarises small amplitude waves in the context of plasma kinetic theory, using the Vlasov description of plasma kinetics coupled with Maxwellian equilibria. Excellent and accessible treatments of the full kinetic theory can be found in a number of texts, for example [15, 19, 23, 50, 52, 67, 88, 90]; only the essential results will be stated here.

7.4.1 Dielectric Function for an Unmagnetized Plasma

For an unmagnetized (and therefore isotropic) kinetic plasma, the general expression for the dielectric function $\epsilon(k, w)$ is

$$\epsilon(k, \omega) = 1 + \frac{q_s^2}{\epsilon_0 m_s k} \int_{-\infty}^{\infty} \frac{\partial f_s / \partial u}{\omega - ku} du \quad \text{(SI)} \qquad (7.135)$$

$$= 1 + \frac{4\pi q_s^2}{m_s k} \int_{-\infty}^{\infty} \frac{\partial f_s / \partial u}{\omega - ku} du \quad \text{(cgs)} \qquad (7.136)$$

where ω is the wave frequency, k the wavenumber, and f_s is the plasma distribution function for species s. The consequent dispersion relation for small amplitude waves is the root of (7.135):

$$1 = \frac{q_s^2}{\epsilon_0 m_s k^2} \int_{-\infty}^{\infty} \frac{\partial f_s / \partial u}{u - \omega/k} \quad \text{(SI)} \qquad (7.137)$$

$$= \frac{4\pi q_s^2}{m_s k^2} \int_{-\infty}^{\infty} \frac{\partial f_s / \partial u}{u - \omega/k} \quad \text{(cgs)} \qquad (7.138)$$

7.4.2 Langmuir Waves

The dispersion relation for high frequency waves in an unmagnetised plasma such that the perturbed distribution function varies as $\exp[i(\boldsymbol{k} \cdot \boldsymbol{r} - \omega t)]$ is

$$\omega = \omega_r + i\omega_i \qquad (7.139)$$

$$\omega_r \approx \omega_{pe} \left(1 + \frac{3}{2} k^2 \lambda_D^2 \right) \qquad (7.140)$$

$$\omega_i = - \left(\frac{\pi}{8} \right)^{1/2} \frac{\omega_{pe}}{k^3 \lambda_D^3} \exp \left(\frac{3}{2} - \frac{1}{2k^2 \lambda_D^2} \right) \qquad (7.141)$$

Restrictions

- the plasma is field-free in equilibrium

- only the electrons are mobile; the ions are stationary

- the equilibrium distribution function for electrons is Maxwellian

- $\dfrac{\omega}{k} \gg c_{th}$

- $k\lambda_D \ll 1$

Note that these waves are purely electrostatic, damped because $\omega_i < 0$, a phenomenon termed Landau damping. These waves are known variously as Langmuir waves, plasma oscillations, electrostatic waves and Langmuir oscillations. These waves have the same frequency for all wavelengths. For $T_e = 0$, we recover the undamped, localised cold plasma oscillation of Section 7.2.4.1.

7.4.3   Ion-Acoustic Waves

Allowing the ions to have a temperature T_i, a low frequency electrostatic wave in which ion motion is important is the ion-acoustic mode, with dispersion relation

$$\omega = \omega_r + i\omega_i \tag{7.142}$$

$$\omega_r = \left(\frac{k_B T_e}{m_i}\frac{k^2}{1 + k^2\lambda_D^2}\right)^{1/2} \tag{7.143}$$

$$\omega_i = -\left(\frac{\pi}{8}\right)^{1/2}\frac{|\omega_r|}{(1 + k^2\lambda_D^2)^{3/2}}\left[\left(\frac{T_e}{T_i}\right)^{3/2}\exp\left(-\frac{T_e/T_i}{2(1 + k^2\lambda_D^2)}\right) + \left(\frac{m_e}{m_i}\right)^{1/2}\right] \tag{7.144}$$

Restrictions

- the plasma is field-free in equilibrium

- derivation of (7.143) and (7.144) requires $(k_B T_i/m_i)^{1/2} < \omega/k < (k_B T_e/m_e)^{1/2}$

- $|\omega_i/\omega_r| \ll 1$ only if $T_i \ll T_e$; if $T_i \gtrsim T_e$ then the wave is heavily damped, and is not a true plasma mode

In the particular case of $T_i \ll T_e$, the frequency and damping terms simplify to

$$\omega_r \approx k \left(\frac{k_B T_e}{m_i}\right)^{1/2} \left(1 + k^2 \lambda_D^2\right)^{-1/2} \tag{7.145}$$

$$\omega_i \approx \omega_r \left(\frac{\pi m_e}{8 m_i}\right)^{1/2} \left(1 + k^2 \lambda_D^2\right)^{-3/2} \tag{7.146}$$

Note that if $k\lambda_D \ll 1$, then there is a common propagation speed for all waves, namely $(k_B T_e/m_i)^{1/2}$ - the ion acoustic speed.

7.4.4 Dielectric Tensor for a Hot Plasma

Using the linearised Vlasov equation for a magnetised plasma,

$$\frac{\partial f_s}{\partial t} + \boldsymbol{u} \cdot \frac{\partial f_s}{\partial \boldsymbol{r}} + \frac{q_s}{m_s}(\boldsymbol{u} \times \boldsymbol{B}_0) \cdot \frac{\partial f_s}{\partial \boldsymbol{u}} + \frac{q_s}{m_s}(\boldsymbol{E} + \boldsymbol{u} \times \boldsymbol{B}) \cdot \frac{\partial f_{s0}}{\partial \boldsymbol{u}} = 0 \quad \text{(SI)} \tag{7.147}$$

$$\frac{\partial f_s}{\partial t} + \boldsymbol{u} \cdot \frac{\partial f_s}{\partial \boldsymbol{r}} + \frac{q_s}{cm_s}(\boldsymbol{u} \times \boldsymbol{B}_0) \cdot \frac{\partial f_s}{\partial \boldsymbol{u}} + \frac{q_s}{m_s}(\boldsymbol{E} + \boldsymbol{u} \times \boldsymbol{B}/c) \cdot \frac{\partial f_{s0}}{\partial \boldsymbol{u}} = 0 \quad \text{(cgs)} \tag{7.148}$$

where f_s is the perturbed distribution function for species s, and f_{s0} is the corresponding equilibrium distribution function, the vector equation for small-amplitude electromagnetic waves in the hot plasma may be written in the form

$$\boldsymbol{k} \times (\boldsymbol{k} \times \boldsymbol{E}) + \frac{\omega^2}{c^2}\boldsymbol{\epsilon}\boldsymbol{E} = 0 \tag{7.149}$$

Take the homogeneous equilibrium magnetic field to lie in the z-direction, and assume that all perturbations have the harmonic form

$$\exp[\mathrm{i}(\boldsymbol{k} \cdot \boldsymbol{r} - \omega t)] \tag{7.150}$$

with

$$\boldsymbol{k} = \hat{\boldsymbol{x}}\,k_\perp + \hat{\boldsymbol{z}}\,k_\| \tag{7.151}$$

Then the dielectric tensor $\boldsymbol{\epsilon}$ can be written in the form

$$\boldsymbol{\epsilon} = \begin{bmatrix} \epsilon_{xx} & \epsilon_{xy} & \epsilon_{xz} \\ \epsilon_{yx} & \epsilon_{yy} & \epsilon_{yz} \\ \epsilon_{zx} & \epsilon_{zy} & \epsilon_{zz} \end{bmatrix} \tag{7.152}$$

where

$$\epsilon_{xx} = 1 + \sum_s \frac{\omega_{ps}^2}{\omega^2} \frac{e^{-b_s}}{b_s} \zeta_{0s} \sum_{n=-\infty}^{\infty} n^2 I_n(b_s) \mathcal{Z}(\zeta_{ns}) \tag{7.153}$$

$$\epsilon_{yy} = 1 + \sum_s \frac{\omega_{ps}^2}{\omega^2} \frac{e^{-b_s}}{b_s} \zeta_{0s} \sum_{n=-\infty}^{\infty} \left\{ n^2 I_n(b_s) + 2b_s^2 \left[I_n(b_s) - I_n'(b_s) \right] \right\} \mathcal{Z}(\zeta_{ns}) \tag{7.154}$$

$$\epsilon_{zz} = 1 - \sum_s \frac{\omega_{ps}^2}{\omega^2} e^{-b_s} \zeta_{0s} \sum_{n=-\infty}^{\infty} I_n(b_s) \zeta_{ns} \mathcal{Z}'(\zeta_{ns}) \tag{7.155}$$

$$\epsilon_{xy} = i \sum_s \pm \frac{\omega_{ps}^2}{\omega^2} e^{-b_s} \zeta_{0s} \sum_{n=-\infty}^{\infty} n \left[I_n(b_s) - I_n'(b_s) \right] \mathcal{Z}(\zeta_{ns}) \tag{7.156}$$

$$\epsilon_{yx} = \epsilon_{xy} \tag{7.157}$$

$$\epsilon_{xz} = \sum_s \frac{\omega_{ps}^2}{\omega^2} (2b_s)^{-1/2} e^{-b_s} \zeta_{0s} \sum_{n=-\infty}^{\infty} n I_n(b_s) \mathcal{Z}'(\zeta_{ns}) \tag{7.158}$$

$$\epsilon_{zx} = -\epsilon_{xz} \tag{7.159}$$

$$\epsilon_{yz} = -i \sum_s \pm \frac{\omega_{ps}^2}{\omega^2} (b_s/2)^{1/2} e^{-b_s} \zeta_{0s} \sum_{n=-\infty}^{\infty} \left[I_n(b_s) - I_n'(b_s) \right] \mathcal{Z}'(\zeta_{ns}) \tag{7.160}$$

$$\epsilon_{zy} = -\epsilon_{yz} \tag{7.161}$$

together with the definitions

$$b_s = \frac{k_\perp k_B T_s}{m_s \omega_{cs}^2} \tag{7.162}$$

$$\zeta_{ns} = \frac{\omega + n|\omega_{cs}|}{k_\parallel} \left(\frac{m_s}{2k_B T_s} \right)^{1/2} \tag{7.163}$$

where I_n is the modified Bessel function, and $\mathcal{Z}$ is the plasma dispersion function.

Restrictions

- only the Vlasov equation was used in the construction of the dielectric tensor, and so no collision terms were modelled, and therefore no plasma transport invoked

- (7.153)-(7.161) depend upon Maxwellian equilibria for both ions and electrons, with no equilibrium streaming of any species

- the results apply only to non-relativistic dynamics

- an isotropic temperature is assumed for all species

- there is no equilibrium electric field present

Using (7.149), (7.152) and (7.153)-(7.161) a generalised dispersion relation can be constructed for waves propagating parallel to, and perpendicular to, the equilibrium magnetic field. Particular examples of these waves are quoted in the following two sections:

7.4.4.1 Parallel Propagation

longitudinal modes   Here set $k_\perp = 0$, which simplifies the dielectric tensor components (7.153)-(7.161), since now only non-zero contributions arise from the terms involving $I_1'(0)$. The dispersion relation for the case $E_z \neq 0$ reduces to the longitudinal electrostatic oscillations for an unmagnetised plasma (see Sections 7.4.2 and 7.4.3).

transverse modes   The waves that have $E_z = 0$ are transverse electromagnetic modes, with special dispersion relations according to frequency range:

Alfvén waves   A low frequency mode, $|\omega| \ll \omega_{ci}$ with dispersion relation

$$\omega = \omega_r + i\omega_i \tag{7.164}$$

$$\omega_r^2 = \frac{k_\| c_a^2}{1 + c_a^2/c^2} \tag{7.165}$$

$$\omega_i = -\frac{\omega_{pi}^2}{|k_\| \bar{u}_i|}(1 + c^2/c_a^2)^{-1} \exp\left(-\frac{m_i c_a^2}{2k_B T_i}\frac{\omega_{ci}^2}{\omega_r^2}\right) \tag{7.166}$$

Generally, Alfvén waves are weakly damped for low frequencies.

whistler waves   At intermediate frequencies, $\omega_{ci} \ll \omega \ll \omega_{ce}$, the real frequency ω_r is given by the cold plasma relation (7.66), with the damping term given by

$$\omega_i \approx -\frac{\omega_{pe}^2}{|k_\||(2k_B T_i/m_i)^{1/2}}\left(1 + k_\| c^2/\omega_r^2\right)^{-1}\exp\left(-\frac{m_e\omega_{ce}^2}{2k_B T_e k_\|^2}\right) \tag{7.167}$$

which is again weakly damped, except for very short wavelengths such that

$$-\frac{m_e\omega_{ce}^2}{2k_B T_e k_\|^2} \approx 1 \tag{7.168}$$

cyclotron waves   For high frequency waves near ω_{ce}, then the dispersion relation takes the local form

$$\omega^2 \approx k_\parallel^2 c^2 - \mathrm{i}\left(\frac{m_e c^2}{2\pi k_B T_e}\right)^{1/2}\omega_{pe}^2 \exp\left[\frac{m_e(\omega-\omega_{ce})^2}{2k_B T_e k_\parallel^2}\right] \tag{7.169}$$

Assuming that $\omega_r \approx \omega_{ce}$, (7.169) can be approximated to yield

$$\omega \approx k_\parallel c - \mathrm{i}\left(\frac{m_e c^2}{8\pi k_B T_e}\right)^{1/2}\frac{\omega_{pe}^2}{\omega_{ce}} \tag{7.170}$$

showing that cyclotron waves propagating parallel to the magnetic field can be strongly damped.

7.4.4.2 Perpendicular Propagation   Here set $k_\parallel = 0$. There are 3 distinct types of wave mode possible.

Ordinary Mode (O-Mode)   The Ordinary mode is a transverse electromagnetic wave, with an electric field component aligned with the equilibrium magnetic field. The dispersion relation corresponds to solutions of $\epsilon_{zz} = k_\perp^2 c^2/\omega^2$, and is

$$\frac{k_\perp^2 c^2}{\omega^2} = 1 - \sum_s \frac{\omega_{ps}^2}{\omega}e^{-b_s}\sum_{n=-\infty}^{\infty}\frac{I_n(b_s)}{\omega-n\omega_{cs}} \tag{7.171}$$

Where only $n=0$ contributions are retained, the dispersion relation simplifies to the cold plasma one (cf (7.68)):

$$\frac{k^2 c^2}{\omega^2} \approx 1 - \frac{\omega_p^2}{\omega^2} \tag{7.172}$$

In addition (7.171) also has solutions for frequencies near the cyclotron frequencies:

$$\omega \approx n\omega_{cs} \tag{7.173}$$

Extraordinary Mode (X-Mode)   The X-mode has an electric field perpendicular to the equilibrium magnetic field, with a small component parallel to $\boldsymbol{k}$; hence it is almost a purely transverse electromagnetic wave, satisfying the dispersion relation

$$\left[1 - \frac{k_\perp^2 c^2}{\omega^2} - \sum_s \frac{\omega_{ps}^2}{\omega} \frac{e^{-b_s}}{b_s} \sum_{n=-\infty}^{\infty} \frac{n^2 I_n(b_s) + 2b_s^2 \left(I_n(b_s) - I_n'(b_s)\right)}{\omega - n\omega_{cs}}\right]$$

$$\times \left[1 - \sum_s \frac{\omega_{ps}^2}{\omega} \frac{e^{-b_s}}{b_s} \sum_{n=-\infty}^{\infty} \frac{n^2 I_n(b_s)}{\omega - n\omega_{cs}}\right]$$

$$= \left[\sum_s \frac{\omega_{ps}}{\omega} e^{-b_s} \sum_{n=-\infty}^{\infty} \frac{n \left(I_n(b_s) - I_n'(b_s)\right)}{\omega - n\omega_{cs}}\right]^2 \qquad (7.174)$$

In the limit of $b_s \to 0$, (7.174) recovers the cold plasma X-mode dispersion relation (7.69).

Bernstein Modes A longitudinal mode, with $E_y = 0, E_x \neq 0$, is another wave solution, satisfying the dispersion relation

$$k_\perp^2 = \sum_s \sum_{n=1}^{\infty} \frac{2n^2 \omega_{ps}^2 \omega_{cs}^2}{\omega^2 - n^2 \omega_{cs}^2} \frac{m_s}{k_B T_s} e^{-b_s} I_n(b_s) \qquad (7.175)$$

Bernstein modes are electrostatic waves propagating perpendicular to the magnetic field. The solutions are separated from one another by band gaps where no propagation is possible. There are simplifying approximations which make (7.175) more tractable:

long wavelength or low temperature If

$$b_e \ll 1 \qquad (7.176)$$

then b_e can be used as an expansion parameter, revealing that the Bernstein modes occur close to cyclotron frequencies, except for the first one:

$$\omega_1 = (\omega_{pe}^2 + \omega_{ce}^2)^{1/2} \qquad (7.177)$$

$$\omega_n \approx n\omega_{ce} \qquad \qquad n \geq 2 \qquad (7.178)$$

very low plasma density : if

$$\omega_{pe}^2 \ll \omega_{ce}^2 \qquad (7.179)$$

the Bernstein modes occur at frequencies close to cyclotron harmonics [50]:

$$\omega^2 = n^2 \omega_{ce}^2 (1 + \alpha_n) \qquad (7.180)$$

$$\alpha_n = 2\frac{\omega_{pe}^2}{\omega_{ce}^2} \frac{e^{-b_e}}{b_e} I_n(b_e) \qquad (7.181)$$

Note also that Bernstein modes are undamped. The existence of these narrow band modes contradicts the cold fluid plasma theory, which has no modes possible between the hybrid resonances (see Section 7.2.4.4).

8
Flows

8.1 NOTATION

Symbol	Meaning	Ref
a	beam radius	
$\boldsymbol{B}$	magnetic flux density	
d	distance between plates	
e	internal energy	(8.126)
E	electric field	
f_n	fractional electrostatic neutralisation	
$\mathcal{F}_\rho$	mass flux through a shock	(8.137)
h	shock strength parameter	(8.136)
$\mathcal{H}_a$	Hartmann number	(2.38)
I_A	Alfvén current	(8.92)
I_0	fundamental current in I_A	(8.95)
$\boldsymbol{J}$	current density	
K	generalized beam perveance	(8.99)
m_s	mass of particle of species s	
$\mathcal{M}$	Mach number	(2.42)
p	gas pressure	
Q	ionization rate	
$\mathcal{R}_p$	pressure ratio across a shock front	(8.129)
$\mathcal{R}_\rho$	density ratio across a shock front	(8.130)
$\bar{u}$	bulk or mean velocity	
γ	polytropic index	
Γ_c	fluid circulation	(8.5)
η	fluid plasma resistivity	
η_v	plasma viscosity	
ν_B	Budker parameter	(8.87)
ρ	mass density of single-fluid plasma	(7.80)
σ_j	square of sound over Alfvén speed on either side of shock	(8.132)
ϕ_i	ionization potential	
ω_p	circular plasma frequency	(2.6)
$\boldsymbol{\omega}$	fluid vorticity	(8.6)

8.2 FUNDAMENTAL RESULTS

8.2.1 Alfvén's Theorem

In a perfectly conducting plasma, the magnetic field lines are transported with any transverse motion of the plasma; the magnetic field behaves as though 'frozen' into the plasma [6]. Hence for a perfectly conducting plasma, the magnetic flux through any closed loop following the material motion remains constant in time. This is expressed mathematically as

$$\frac{\partial \Phi}{\partial t} + \boldsymbol{u} \cdot \nabla \Phi = \oint (\boldsymbol{E} + \boldsymbol{u} \times \boldsymbol{B}) \, \mathrm{d}l \qquad \text{(SI)} \qquad (8.1)$$

$$= \oint (\boldsymbol{E} + \boldsymbol{u} \times \boldsymbol{B}/c) \, \mathrm{d}l \qquad \text{(cgs)} \qquad (8.2)$$

$$= 0 \qquad \qquad \text{if } \eta = 0 \qquad (8.3)$$

Consequently, all the fluid particles which initially lie on a particular field line continue to do so.

8.2.2   Cowling's Anti-Dynamo Theorem

A steady axisymmetric magnetic field configuration cannot be self-maintained by the current it induces in a plasma undergoing steady motion about the axis of symmetry [25], [26]. In other words, no steady, axially symmetric dynamo is possible [8].

8.2.3   Ferraro's Law of Isorotation

Consider an axisymmetric steady MHD plasma flow in cylindrical co-ordinates, in which the azimuthal component of magnetic field is zero and the plasma resistivity is a constant. Then the local angular velocity of a fluid element does not vary along a field line; in other words, the fluid at all points on a field line rotates about the axis at a uniform angular velocity $\Omega_u = u_\theta / r$:

$$(\boldsymbol{B} \cdot \nabla)\Omega_u = 0 \qquad (8.4)$$

where u_θ is the azimuthal component of the plasma velocity. This result holds for finite resistivity plasmas.

8.2.4   Kelvin's Vorticity Theorem

The circulation Γ_c round a closed loop moving with the fluid is defined as

$$\Gamma_c = \oint \boldsymbol{u} \cdot \mathrm{d}l \qquad (8.5)$$

It is related to the fluid vorticity $\boldsymbol{\omega}$,

$$\boldsymbol{\omega} = \nabla \times \boldsymbol{u} \qquad (8.6)$$

by

$$\Gamma_c = \iint \boldsymbol{\omega} \cdot \mathrm{d}\boldsymbol{a} \qquad (8.7)$$

where the integration is taken across any surface spanning the loop. If $\Gamma_c = 0$ around all loops, then $\boldsymbol{\omega} = 0$ everywhere, and vice-versa (provided that the fluid domain is simply connected, that is, there are no obstacles that exclude the plasma from a finite region).

Kelvin's theorem states that the rate of change of Γ_c is zero only if the force per unit mass is irrotational:

$$\frac{\partial \Gamma_c}{\partial t} + (\boldsymbol{u} \cdot \nabla)\Gamma_c = \iint \nabla \times (\boldsymbol{F}/\rho) \cdot \mathrm{d}\boldsymbol{a} \tag{8.8}$$

where $\boldsymbol{F}$ is the body force, ρ is the fluid mass-density, and the integration is once more taken across any surface spanning the loop.

8.3 HYDROMAGNETIC FLOWS

The hydromagnetic Navier-Stokes equation is the momentum equation for a viscous magnetofluid:

$$\rho\left(\frac{\partial \boldsymbol{u}}{\partial t} + \boldsymbol{u} \cdot \nabla\boldsymbol{u}\right) = -\nabla p + \boldsymbol{J} \times \boldsymbol{B} + \frac{\eta_v}{3}\nabla(\nabla \cdot \boldsymbol{u}) + \eta_v\nabla^2\boldsymbol{u} \qquad \text{(SI)} \quad (8.9)$$

$$\rho\left(\frac{\partial \boldsymbol{u}}{\partial t} + \boldsymbol{u} \cdot \nabla\boldsymbol{u}\right) = -\nabla p + \frac{1}{c}\boldsymbol{J} \times \boldsymbol{B} + \frac{\eta_v}{3}\nabla(\nabla \cdot \boldsymbol{u}) + \eta_v\nabla^2\boldsymbol{u} \qquad \text{(cgs)}$$
$$\tag{8.10}$$

where the viscosity η_v is assumed constant.

Taken together with the density equation,

$$\frac{\partial \rho}{\partial t} + \nabla \cdot (\rho\boldsymbol{u}) \tag{8.11}$$

and the reduced Maxwell equations,

$$\nabla \times \boldsymbol{E} = -\frac{\partial \boldsymbol{B}}{\partial t} \qquad \text{(SI)} \tag{8.12}$$

$$\nabla \times \boldsymbol{E} = -\frac{1}{c}\frac{\partial \boldsymbol{B}}{\partial t} \qquad \text{(cgs)} \tag{8.13}$$

$$\nabla \times \boldsymbol{B} = \mu_0\boldsymbol{J} \qquad \text{(SI)} \tag{8.14}$$

$$\nabla \times \boldsymbol{B} = \frac{4\pi}{c}\boldsymbol{J} \qquad \text{(cgs)} \tag{8.15}$$

$$\boldsymbol{E} + \boldsymbol{u} \times \boldsymbol{B} = \eta\boldsymbol{J} \qquad \text{(SI)} \tag{8.16}$$

$$\boldsymbol{E} + \boldsymbol{u} \times \boldsymbol{B}/c = \eta\boldsymbol{J} \qquad \text{(cgs)} \tag{8.17}$$

these relations form a complete set for the study of hydromagnetic flows.

The curl of the Navier-Stokes equation (8.9) yields the relation for the evolution of vorticity in an incompressible, uniform density magnetofluid:

$$\frac{\partial \boldsymbol{\omega}}{\partial t} + (\boldsymbol{u} \cdot \nabla)\boldsymbol{\omega} - \boldsymbol{\omega} \cdot \nabla \boldsymbol{u} = \nabla \times (\boldsymbol{J} \times \boldsymbol{B}/\rho) \qquad \text{(SI)} \qquad (8.18)$$

$$= \nabla \times (\boldsymbol{J} \times \boldsymbol{B}/(c\rho)) \qquad \text{(cgs)} \qquad (8.19)$$

The term $\boldsymbol{\omega} \cdot \nabla \boldsymbol{u}$ is referred to as the vorticity stretching term [84].

8.3.1 Hartmann Flow

The steady (that is, time independent) flow of a magnetofluid along a duct of constant rectangular cross-section across a uniform applied magnetic field is termed Hartmann Flow. In cartesian co-ordinates, take the duct height to be $2d$, (that is, $-d \leq z \leq d$), and assume that its width (y-direction) $W \gg d$ and length (in x-direction) $L \gg d$ permit the problem to be treated as a 1-dimensional flow between two infinite parallel plates.

Assume a flow $\boldsymbol{u}$ in the x-direction along the duct, with an applied magnetic field B_0 in the z-direction. Then we have

$$\boldsymbol{u} = \hat{\mathbf{x}}\, u(z) \qquad (8.20)$$

$$\boldsymbol{B} = \hat{\mathbf{x}}\, B_x + \hat{\mathbf{z}}\, B_0 \qquad (8.21)$$

$$\boldsymbol{E} = \hat{\mathbf{y}}\, E_0 \qquad (8.22)$$

$$\boldsymbol{J} = \hat{\mathbf{y}}\, J_y \qquad (8.23)$$

$$\frac{\partial p}{\partial x} = p_{x0} \qquad (8.24)$$

where E_0 and p_{x0} are constants. The boundary conditions are

$$u(\pm d) = 0 \qquad (8.25)$$

The solution for the flow in the channel is given by

$$u(z) = d^2 \frac{p_{x0} + B_0 E_0/\eta}{\eta_v \mathcal{H}_a^2} \left[1 - \frac{\cosh(\mathcal{H}_a z/d)}{\cosh \mathcal{H}_a} \right] \qquad \text{(SI)} \qquad (8.26)$$

$$= d^2 \frac{p_{x0} + B_0 E_0/(c\eta)}{\eta_v \mathcal{H}_a^2} \left[1 - \frac{\cosh(\mathcal{H}_a z/d)}{\cosh \mathcal{H}_a} \right] \qquad \text{(cgs)} \qquad (8.27)$$

The Hartmann number, $\mathcal{H}_a$, is defined by

$$\mathcal{H}_a = \frac{B_0 d}{(\eta \eta_v)^{1/2}} \qquad \text{(SI)} \qquad (8.28)$$

$$\mathcal{H}_a = \frac{B_0 d}{c\,(\eta \eta_v)^{1/2}} \qquad \text{(cgs)} \qquad (8.29)$$

and $\bar{u}$ is the mean speed:

$$\bar{u} = \frac{1}{2d} \int_{-d}^{d} u(z')\mathrm{d}z' \tag{8.30}$$

$$= 3v_{Po} \left(1 + \frac{B_0 E_0}{P_0 \eta_v}\right) (\mathcal{H}_a - \tanh \mathcal{H}_a) \qquad \text{(SI)} \tag{8.31}$$

$$= 3v_{Po} \left(1 + \frac{B_0 E_0}{c P_0 \eta_v}\right) (\mathcal{H}_a - \tanh \mathcal{H}_a) \qquad \text{(cgs)} \tag{8.32}$$

in which v_{Po} is the mean speed for Poiseuille flow in the same rectangular cross-section duct:

$$v_{Po} = \frac{d^2 p_{x0}}{3\eta_v} \tag{8.33}$$

The current density in the duct is

$$J_y = \eta \left(E_0 - u B_0\right) \qquad \text{(SI)} \tag{8.34}$$

$$= \eta \left(E_0 - u B_0/c\right) \qquad \text{(cgs)} \tag{8.35}$$

with a mean current density given by

$$\bar{J} = \frac{1}{2d} \int_{-d}^{d} J_y \mathrm{d}z \tag{8.36}$$

$$= \frac{1}{\eta}(E_0 - \bar{u} B_0) \qquad \text{(SI)} \tag{8.37}$$

$$= \frac{1}{\eta}(E_0 - \bar{u} B_0/c) \qquad \text{(cgs)} \tag{8.38}$$

The induced magnetic field is given by

$$B_x = \frac{\mu_0 d \left(p_{x0} + B_0 \bar{J}\right)}{B_0 \sinh \mathcal{H}_a} \sinh \left(\mathcal{H}_a z/d\right) - \frac{\mu_0 p_{x0} z}{B_0} \qquad \text{(SI)} \tag{8.39}$$

$$B_x = \frac{4\pi d \left(c p_{x0} + B_0 \bar{J}\right)}{c B_0 \sinh \mathcal{H}_a} \sinh \left(\mathcal{H}_a z/d\right) - \frac{4\pi p_{x0} z}{B_0} \qquad \text{(cgs)} \tag{8.40}$$

Finally, the pressure in the duct is

$$p(x, z) = \text{constant} - p_{x0} x - \frac{B_x^2}{2\mu_0} \qquad \text{(SI)} \tag{8.41}$$

$$= \text{constant} - p_{x0} x - \frac{B_x^2}{8\pi} \qquad \text{(cgs)} \tag{8.42}$$

The significance of these results lies in the sign of the mean current density. If $\bar{J} B_0 < 0$ then fluid flow is opposed by magnetic forces, and electrical power is extracted from the fluid (the MHD generator); conversely, if $\bar{J} B_0 > 0$ then the fluid is accelerated by the magnetic forces (the plasma pump).

8.3.2 Couette Flow

The flow between two parallel plates, one of which is moving with respect to the other, is termed Couette flow. The problem is very similar to the Hartmann flow problem of 8.3.1, and so the same geometry and notation is used here, but since the flow is induced by the viscous drag of the plate at $z = +d$ moving with speed u_0 in the x-direction, the velocity boundary conditions for the flow are

$$u(-d) = 0 \tag{8.43}$$

$$u(+d) = u_0 \tag{8.44}$$

The solution for the flow speed between the plates is then

$$u(z) = u_0 \frac{\sinh\left(\mathcal{H}_a z/d\right)}{\sinh\left(2\mathcal{H}_a\right)} + 2\frac{E_0}{B_0} \frac{\sinh\left(\mathcal{H}_a z/(2d)\right) \sinh\left[\mathcal{H}_a(1 - z/(2d))\right]}{\cosh \mathcal{H}_a} \quad \text{(SI)}$$

$$\tag{8.45}$$

$$= u_0 \frac{\sinh\left(\mathcal{H}_a z/d\right)}{\sinh\left(2\mathcal{H}_a\right)} + 2\frac{cE_0}{B_0} \frac{\sinh\left(\mathcal{H}_a z/(2d)\right) \sinh\left[\mathcal{H}_a(1 - z/(2d))\right]}{\cosh \mathcal{H}_a} \quad \text{(cgs)}$$

$$\tag{8.46}$$

8.3.3 Field-Aligned Flows

The equations governing an incompressible, viscous, resistive MHD plasma are as follows [15, 48, 84]:

$$\nabla \cdot \boldsymbol{u} = 0 \tag{8.47}$$

$$\boldsymbol{u} \cdot \nabla \rho = 0 \tag{8.48}$$

$$\rho\left(\frac{\partial \boldsymbol{u}}{\partial t} + (\boldsymbol{u} \cdot \nabla)\boldsymbol{u}\right) = -\nabla p + \boldsymbol{J} \times \boldsymbol{B} + \eta_v \nabla^2 \boldsymbol{u} \quad \text{(SI)} \tag{8.49}$$

$$= -\nabla p + \boldsymbol{J} \times \boldsymbol{B}/c + \eta_v \nabla^2 \boldsymbol{u} \quad \text{(cgs)} \tag{8.50}$$

$$\frac{\partial \boldsymbol{B}}{\partial t} = \nabla \times (\boldsymbol{u} \times \boldsymbol{B}) - \nabla \times (\eta \nabla \times \boldsymbol{B}) \quad \text{(SI)} \tag{8.51}$$

$$= \nabla \times (\boldsymbol{u} \times \boldsymbol{B}) - c\nabla \times (\eta \nabla \times \boldsymbol{B}) \quad \text{(cgs)} \tag{8.52}$$

where (8.47) ensures incompressibility and (8.48) requires the density to be constant on a streamline.

Taking the magnetic field to be aligned with the velocity field,

$$\boldsymbol{B} = \lambda \boldsymbol{u} \tag{8.53}$$

for some scalar function λ yields immediately

$$\boldsymbol{u} \cdot \nabla \lambda = 0 \tag{8.54}$$

that is, λ must be a constant on a streamline.

Then any steady field aligned flow must satisfy [48]

$$\nabla \times (\eta \nabla \times \boldsymbol{B}) = 0 \qquad (8.55)$$

In addition, the equation of motion for steady flows yields

$$\tilde{\rho}\boldsymbol{u} \cdot \boldsymbol{u} = -\nabla \tilde{p} + \eta_v \nabla^2 \boldsymbol{u} \qquad (8.56)$$

where

$$\tilde{\rho} = \rho - \lambda^2/\mu_0 \qquad \text{(SI)} \qquad (8.57)$$

$$= \rho - \lambda^2/(4\pi) \qquad \text{(cgs)} \qquad (8.58)$$

$$\tilde{p} = p + \lambda^2 u^2/(2\mu_0) \qquad \text{(SI)} \qquad (8.59)$$

$$= p + \lambda^2 u^2/(8\pi) \qquad \text{(cgs)} \qquad (8.60)$$

are the modified mass density and pressure. Note that since (8.56) is analogous to the form of Navier-Stokes equation for incompressible hydrodynamics, a perfect analogy depends upon solving (8.55) in a manner compatible with hydrodynamics.

8.3.3.1 η, λ Constant:
Here, the flow is potential, satisfying

$$\boldsymbol{u} = -\nabla \phi \qquad (8.61)$$

$$\boldsymbol{B} = -\lambda \nabla \phi \qquad (8.62)$$

$$\nabla^2 \phi = 0 \qquad (8.63)$$

where ϕ is the velocity potential for steady, incompressible and irrotational flow. Hence an arbitrary potential flow of a viscous, incompressible fluid in the absence of a magnetic field provides a solution to an MHD parallel steady flow, if η and λ are constants.

Note also that in the steady flow for constant density ρ,

$$\tilde{\rho}\nabla \times (\boldsymbol{\omega} \times \boldsymbol{u}) = \tilde{\rho}\left[\frac{\partial \boldsymbol{\omega}}{\partial t} + (\boldsymbol{u} \cdot \nabla)\boldsymbol{\omega} - \boldsymbol{\omega} \cdot \nabla \boldsymbol{u}\right] = \nabla \times \mathcal{F} \qquad (8.64)$$

where $\mathcal{F}$ is any rotational body force. The significance of $\tilde{\rho}$ is clear from (8.64) and (8.18), since if $\lambda^2/\mu_0 > \rho$, then $\tilde{\rho} < 0$ and the fluid vorticity increases in the *opposite* direction to the rotationality of $\mathcal{F}$, an effect termed *negative inertia* [84].

Certain special cases are discussed in the following sub-sections; a more general treatment can be found in [38], and the references therein.

8.3.3.2 $\eta = 0$
For a perfectly conducting plasma, (8.55) is automatically solved in field-aligned flows. Given that (8.48) and (8.54) hold, then any classical hydrodynamical potential flow for which $\tilde{\rho}$ is constant everywhere can be mapped to an incompressible ideal MHD flow, which is not current-free if λ is not a constant.

8.3.3.3 Inviscid Flows   In the particular case of $\eta_v = 0$, Bernoulli's equation holds along a streamline:

$$\frac{1}{2}\tilde{\rho}u^2 + \tilde{p} = \frac{1}{2}\rho u^2 + p = \text{constant} \tag{8.65}$$

8.4   SOLAR WIND

Classical solutions to the solar wind are strictly hydrodynamical in nature, or equivalently force-free, in that the magnetic force term $\boldsymbol{J} \times \boldsymbol{B}$ does not enter into the equilibrium equations. The basic classical model of dynamical equilibrium is due to Parker, which assumes an isothermal, spherically symmetric wind, for which the velocity $\boldsymbol{u} = \hat{\boldsymbol{r}}\, u$ is given by

$$(\mathcal{M}^2 - 1) - \ln\left(\mathcal{M}^2\right) = 4\ln\left(\frac{r}{r_c}\right) + \frac{2GM_\odot}{u_{th}^2}\left(\frac{1}{r} - \frac{1}{r_c}\right) \tag{8.66}$$

where $\mathcal{M} = u/c_{th}$ is the Mach number, c_{th} is the (constant) gas sound speed, G is the gravitational constant, and $M_\odot$ is the solar mass. The parameter r_c is the distance at which the wind speed becomes supersonic, that is, $\mathcal{M} = 1$. This critical point may be inverted to yield the temperature:

$$T = \frac{GM_\odot}{r_c} m_p 4k_B \leq 6 \times 10^6\,\text{K} \tag{8.67}$$

Note that the Parker model has several restrictions which are not appropriate to the solar wind:

- the wind is isothermal, which is in conflict with the actual measurements;

- the wind is spherically symmetric, which is not true;

- the magnetic field plays no part in determining the equilibrium.

Extensions to the theoretical and numerical modelling addressing the above points are discussed in [14, 56, 73]. Typical data for the solar wind [56] are given in Table 8.2, in which solar parameters and energy densities are compared at various positions in the solar wind. The quantities are defined as follows:

n	solar wind particle number density
ρ	solar wind mass density
r	distance from sun
$R_\odot$	solar radius
n_e	electron number density
T	solar wind gas temperature
B	solar wind magnetic flux density
v	solar wind bulk gas speed

$$E_v = \tfrac{1}{2}\rho v^2 \qquad \text{kinetic energy density} \tag{8.68}$$

$$E_T = \tfrac{3}{2}nk_B T \qquad \text{thermal energy density} \tag{8.69}$$

$$E_G = GM\rho/r \qquad \text{gravitational potential energy density} \tag{8.70}$$

$$E_M = B^2/(2\mu_0) \qquad \text{magnetic energy density, SI} \tag{8.71}$$

$$= B^2/(8\pi) \qquad \text{magnetic energy density, cgs} \tag{8.72}$$

Table 8.2 Average quiet sun conditions in the solar equatorial plane, *reproduced from [56] with permission* See text for symbol key

$r/R_\odot$	1.03	1.5	3	5	10	215 (1AU)
n_e/m^{-3}	2×10^{14}	2×10^{13}	4×10^{11}	4×10^{10}	4×10^{9}	7×10^{6}
n_e/cm^{-3}	2×10^{8}	2×10^{7}	4×10^{5}	4×10^{4}	4×10^{3}	7
T/K	2×10^{6}	1×10^{6}	7×10^{5}	5×10^{5}	4×10^{5}	4×10^{4}
B/T	10^{-4}	4×10^{-5}	10^{-5}	4×10^{-6}	10^{-6}	3×10^{-9}
B/Gauss	1	0.4	0.1	0.04	0.01	3×10^{-5}
$v/\text{km s}^{-1}$	0.6	3	34	130	280	360
$E_v/\text{J m}^{-3}$	4×10^{4}	8×10^{4}	2×10^{5}	3×10^{5}	2×10^{5}	5×10^{2}
$E_v/\text{eV cm}^{-3}$	4×10^{5}	8×10^{5}	2×10^{6}	3×10^{6}	2×10^{6}	5×10^{3}
$E_T/\text{J m}^{-3}$	9×10^{9}	4×10^{8}	7×10^{6}	5×10^{5}	4×10^{4}	7
$E_T/\text{eV cm}^{-3}$	9×10^{10}	4×10^{9}	7×10^{7}	5×10^{6}	4×10^{5}	70
$E_M/\text{J m}^{-3}$	3×10^{9}	5×10^{8}	3×10^{7}	4×10^{6}	3×10^{5}	2
$E_M/\text{eV cm}^{-3}$	3×10^{10}	5×10^{9}	3×10^{8}	4×10^{7}	3×10^{6}	20
$E_G/\text{J m}^{-3}$	4×10^{10}	2×10^{9}	3×10^{7}	10^{6}	8×10^{4}	6
$E_G/\text{eV cm}^{-3}$	4×10^{11}	2×10^{10}	3×10^{8}	10^{7}	8×10^{5}	60

8.5 NEUTRAL GAS/MAGNETIZED PLASMA FLOWS

A stationary magnetized plasma can be accelerated by a moving neutral gas if the speed of the latter exceeds the Alfvén critical speed v_c for ionization [7, 55, 65] (see also Section 3.5.2), given by

$$\tfrac{1}{2}m_n v_c^2 = e\phi_i \tag{8.73}$$

where m_n is the neutral gas particle mass, and ϕ_i its ionization potential.

The neutral gas passes through an initially stationary magnetized plasma, becoming ionized as it does so. The newly created ions and electrons then become part of the plasma component, adding momentum characteristic of the neutral gas to the plasma. The latter then begins to accelerate, reaching a terminal velocity.

For the one-dimensional treatment in which the neutral gas and plasma have velocities in the x-direction, and the magnetic field lies in the y-direction, the appropriate steady-state magnetofluid equations are [65]

$$\frac{\partial}{\partial x}(\rho v) = Q n_e m_n \tag{8.74}$$

$$\rho v \frac{\partial v}{\partial x} = -\frac{\partial p}{\partial x} + \hat{x} \cdot \boldsymbol{J} \times \boldsymbol{B} - n_e Q m_n(v - v_n) \quad \text{(SI)} \tag{8.75}$$

$$= -\frac{\partial p}{\partial x} + \hat{x} \cdot \boldsymbol{J} \times \boldsymbol{B}/c - n_e Q m_n(v - v_n) \quad \text{(cgs)} \tag{8.76}$$

$$\frac{\partial}{\partial x}(\tfrac{1}{2}\rho \bar{c}^2 v) + p\frac{\partial v}{\partial x} = Q n_e [\tfrac{1}{2}m_n(v - v_n)^2 - e\phi_i] \tag{8.77}$$

$$\boldsymbol{E} + \boldsymbol{v} \times \boldsymbol{B} = (\boldsymbol{J} \times \boldsymbol{B} - \nabla p_e)/(n_e e) \quad \text{(SI)} \tag{8.78}$$

$$= Q n_e [\tfrac{1}{2}m_n(v - v_n)^2 - e\phi_i] \tag{8.79}$$

$$\boldsymbol{E} + \boldsymbol{v} \times \boldsymbol{B}/c = (\boldsymbol{J} \times \boldsymbol{B}/c - \nabla p_e)/(n_e e) \quad \text{(cgs)} \tag{8.80}$$

where: $\rho \approx n_e m_n$ is the plasma mass density; $\boldsymbol{v} = \hat{x}v$ is the plasma velocity; Q is the ionization rate; n_e the electron number density; p the total plasma pressure; p_e the electron pressure; $\boldsymbol{J}, \boldsymbol{B}$ the current density and magnetic field, respectively; $\boldsymbol{v}_n = \hat{x}v_n$ the neutral gas velocity; and $\rho\bar{c}^2 v/2$ the plasma thermal energy, such that $p = (\gamma - 1)\rho\bar{c}^2/2$ where $\gamma = 5/3$ is the adiabatic index.

For the case where the plasma speed is small at $x = -\infty$, and increases monotonically with x, then $v \to v_-$ as $x \to \infty$ such that

$$|v - v_-| \propto \exp\left[-\frac{\gamma + 1}{2}Q \left| \frac{v_+ - v_-}{v_-^2 - c_{th}^2 - c_a^2} \right| \right] \tag{8.81}$$

where

$$v_{\pm} = \frac{v_n}{\gamma + 1} \left\{ \gamma \pm \left[(\gamma^2 - 1) \frac{v_c^2}{v_n^2} + 1 \right]^{\frac{1}{2}} \right\}$$ (8.82)

where c_a, c_{th} are the Alfvén and plasma thermal speeds respectively.

The condition that ionization does not cease before the terminal speed is reached places the following constraint on the neutral gas speed:

$$v_n > \frac{3\gamma - 1}{2[\gamma(\gamma - 1)]^{\frac{1}{2}}} v_c \approx 1.8 v_c$$ (8.83)

That is, the neutral gas speed must be approximately twice the Alfvén critical speed.

Note that in the limit of very large neutral gas speeds,

$$\frac{v_-}{v_n} \approx \frac{\gamma - 1}{\gamma + 1}$$ (8.84)

similar to the condition prevailing for strong shocks (see Section 8.7.3.1).

8.6 BEAMS

A plasma beam [54] is a directed stream of charged particles in which the individual particle motion makes a small angle with the beam axis, and in which the thermal spread in energy of the particles is small compared to their total energy. The motion of particles in a beam depends on the applied external fields, and also on the self-field arising from collective plasma effects. Interactions between beam particles can take two forms: (i) a space-charge force, which creates long-range electric fields and is independent of the particulate nature of the beam; and (ii) short-range collisional forces, in which beam particles interact directly with one another, and also with any background particles.

It is assumed that the beam is sufficiently dense that collective effects are significant. A laminar beam is one in which the velocity distribution at a point is single valued.

8.6.1 Beam Parameters

8.6.1.1 Relativistic Factors The notation β_v and γ_v will denote the relativistic parameters

$$\beta_v = \frac{v}{c}$$ (8.85)

$$\gamma_v = \left(1 - \beta_v^2\right)^{-1/2}$$ (8.86)

where v is the speed of a beam *particle*. In general, β_v and γ_v are functions of position within the beam.

8.6.1.2 Budker Parameter   The Budker parameter ν_B is the product of the number N of charged particles per unit length of a beam, and the classical radius of the particle [54]:

$$\nu_B = \frac{Nq^2}{4\pi\epsilon_0 m_0 c^2} \quad \text{(SI)} \tag{8.87}$$

$$\nu_B = \frac{Nq^2}{m_0 c^2} \quad \text{(cgs)} \tag{8.88}$$

where q is the charge on the particle, and m_0 is the particle rest mass.

If the beam is spatially uniform, with constant number density n,

$$\nu_B = \frac{a^2 \omega_p^2}{4c^2} \tag{8.89}$$

where a is the beam radius, γ is the relativistic factor, and ω_p is the plasma frequency of the (relativistic) beam, given by

$$\omega_p^2 = \frac{nq^2}{\gamma\epsilon_0 m_0} \quad \text{(SI)} \tag{8.90}$$

$$= \frac{4\pi nq^2}{\gamma m} \quad \text{(cgs)} \tag{8.91}$$

8.6.1.3 Neutralization   The neutralization of charged particle beams by particles of the opposite sign is a practical feature of all beams, usually achieved by the ionisation of any residual gas in the vacuum system, or by particles in the background plasma. The fraction of beam particles neutralized in this way will be denoted f_n; in general, f_n is a function of position along the beam.

8.6.1.4 Alfvén Current   The maximum current possible in a collimated cylindrical charged particle beam under the influence of its own magnetic field is given by the Alfvén current [5]

$$I_A = \frac{4\pi\epsilon_0 m_0 c^3}{q}\beta_v\gamma_v \quad \text{(SI)} \tag{8.92}$$

$$I_A = \frac{m_0 c^3 \beta_v \gamma_v}{q} \quad \text{(cgs)} \tag{8.93}$$

where the particles have rest mass m_0 and carry charge q.

Restrictions

- the current density is uniform across the beam

- the particles are mono-energetic and identical

- perfect charge neutralization is provided by oppositely charged static background particles co-spatial with the beam

- the beam therefore has a self magnetic field, but has a constant electric potential throughout

For electrons, (8.92) may be approximated as

$$I_A = I_0 \beta_v \gamma_v \tag{8.94}$$

where

$$I_0 \simeq 17 \text{kA} \qquad \text{(SI)} \tag{8.95}$$

$$\simeq 51 \times 10^{12} \, \text{statamp} \quad \text{(cgs)} \tag{8.96}$$

Electron beams with currents in excess of I_A produce electron trajectories with a drift in the opposite direction to the current, because of the particular form of the magnetic field under these assumptions. In this way, the excess current is either cancelled, or the beam becomes immediately unstable.

An alternative formulation [53] defines I_A by equating the electron Larmor radius for the maximum self magnetic field to the beam radius. Generalizing this to include fractional neutralization only, and assuming a uniform, mono-energetic electron beam, then the current limit can be written as

$$I_A = I_0 \frac{\beta_v^3 \gamma_v}{\beta_v^2 + f_n - 1} \tag{8.97}$$

where f_n is the fractional electrostatic neutralization. This admits arbitrarily large currents in a uniform beam for $f_n \sim 1 - \beta_v^2$.

If in addition to partial electrostatic neutralization, there is also partial magnetic neutralization, where large numbers of the background electrons drift in the opposite direction to the beam current producing a partial cancellation of the beam's magnetic field, then I_A can be modified to [40]

$$I_A = I_0 \frac{\beta_v^3 \gamma_v}{\beta_v^2(1 - f_m) - (1 - f_n)} \tag{8.98}$$

where f_m is the fractional magnetic neutralization.

Further treatment of this problem [13, 40] shows that I_A can be exceeded in three ways: (i) if the current density is concentrated near the edge of the beam, so that beam electrons leave the high magnetic field regions before being turned back on themselves; (ii) if the beam propagates into a high density background plasma, in which plasma currents can be induced which cancel the beam's self-field; and (iii) if a strong axial magnetic guide field B_g is added to the beam in order to limit the radial excursion of beam electrons, such that $B_g \gg |B_{max}|(1 - \beta_v^2 - f_n)/\beta_v^2$, where B_{max} is the peak self-field of the beam.

8.6.1.5 Generalized Perveance The generalized perveance K is the dimensionless net radial force acting on particles in a uniform cylindrical beam with zero externally applied fields, and is defined by

$$K = \frac{2\nu_B}{\beta_v^2 \gamma_v} \left(1 - \beta_v^2 - f_n\right) \tag{8.99}$$

$$= \frac{2\nu_B}{\beta_v^2 \gamma_v} \left(\frac{1}{\gamma_v^2} - f_n\right) \tag{8.100}$$

If $f < \gamma_v^{-2}$ (that is, $K > 0$) the beam spreads radially; if $f > \gamma_v^{-2}$ (that is, $K < 0$) the beam pinches radially inwards.

8.6.2   Special Cases

8.6.2.1 Cylindrical Beam with Zero Applied Magnetic Field A uniform parallel beam of particles is injected into a field-free space.

Restrictions

- the beam is laminar, and collisionless

- the beam is partially charge neutralized by a fraction f of oppositely charged particles with negligible axial motion

- only self-fields are present

- there is no variation with axial co-ordinate

If the current is vanishingly small, the beam continues as a perfect cylinder, since the self-fields are negligible.

For non-trivial beam current, the only forces acting on the beam particles are purely radial, such that

$$r \frac{d^2 r}{dz^2} = K \tag{8.101}$$

where r is the radial co-ordinate of a beam particle, z is the axial co-ordinate, and K is the generalized beam perveance (8.99) which quantifies the competition between the outward electric field and the inward magnetic pinch. As a result, additional focusing or defocusing of the beam will occur according as to whether $K < 0$ or $K > 0$. If $K = -2$ then the beam is magnetically pinched so that the radius of curvature of beam-edge particles is equal to the beam radius. As a consequence, the beam carries the maximum possible current, ie $I = I_A$.

The solution to (8.101) can be written [54]

$$\frac{z}{a_0} = \left(\frac{2}{|K|}\right)^{1/2} \times \begin{cases} \int_0^{\sqrt{\ln(r/a_0)}} \exp(u^2) du & K > 0 \\ \int_0^{\sqrt{\ln(a_0/r)}} \exp(-u^2) du & K < 0 \end{cases} \tag{8.102}$$

where a_0 is that radius at which $dr/dz = 0$.

8.6.2.2 Cylindrical Beam in Infinite Magnetic Field
The infinite magnetic field serves to suppress all transverse motion of the beam particles.

Restrictions

- there is no transverse motion of the particles

- the beam is laminar, and collisionless

- the self magnetic field is irrelevant

- the beam has circular symmetry, with radius a

- there is no variation with axial co-ordinate

- beam particles each have the same rest mass m_0 and carry charge q

- the beam is perfectly neutralized overall, although the charge balance is provided external to the beam, so that there is a net electric potential associated with it

- the beam carries a total current I, and possesses a line charge density N, defined below

$$I = \int_0^a 2\pi q r n(r)\beta_v(r)c\, dr, \tag{8.103}$$

$$N = \int_0^a 2\pi r n(r)dr. \tag{8.104}$$

The electric field associated with the beam is purely radial, arising from the beam charge density:

$$E_r = \begin{cases} \dfrac{q}{\epsilon_0 r}\displaystyle\int_0^r r'n(r')dr' & r \le a \quad \text{(SI)} \\[2ex] \dfrac{4\pi q}{r}\displaystyle\int_0^r r'n(r')dr' & r \le a \quad \text{(cgs)} \\[2ex] \dfrac{Nq}{2\pi\epsilon_0 r} & r \ge a \quad \text{(SI)} \\[2ex] \dfrac{2Nq}{r} & r \ge a \quad \text{(cgs)} \end{cases} \tag{8.105}$$

The beam potential ϕ can be defined by

$$q\phi(a) + (\gamma_v(a) - 1)m_0 c^2 = 0 \tag{8.106}$$

which sets the zero of potential such that electric potential energy of a particle at the beam edge balances the particle's kinetic energy there. In this way, the particle energy as a function of radius can be determined:

$$(\gamma_v - 1)m_0c^2 = (\gamma_v(a) - 1)m_0c^2 - q\int_r^a E_r(r')\mathrm{d}r' \tag{8.107}$$

Uniform Current Density   For the simple case of uniform current density, that is,

$$n(r)\beta_v(r) = \text{constant} \tag{8.108}$$

$$= \frac{I}{\pi a^2 qc} \tag{8.109}$$

it is possible to define the particle speed as a function of radius in the following way:

$$\frac{1}{r}\frac{\mathrm{d}}{\mathrm{d}r}\left(r\frac{\mathrm{d}\gamma_v}{\mathrm{d}r}\right) = \frac{qI}{\pi a^2 \epsilon_0 m_0 c^3 \beta_v} \quad \text{(SI)} \tag{8.110}$$

$$= \frac{4qI}{a^2 m_0 c^3 \beta_v} \quad \text{(cgs)} \tag{8.111}$$

There are two special cases which admit simple solutions to (8.110):

1. *Non-Relativistic Particles*   Using (8.107) and (8.105) yields

$$\beta_v(r) = \left(\frac{9qI}{8\pi\epsilon_0 m_0 c^3}\right)^{1/3}\left(\frac{r}{a}\right)^{2/3} \quad \text{(SI)} \tag{8.112}$$

$$= \left(\frac{9qI}{2m_0 c^3}\right)^{1/3}\left(\frac{r}{a}\right)^{2/3} \quad \text{(cgs)} \tag{8.113}$$

$$\nu_B = \tfrac{1}{3}\beta_v^2(a) \tag{8.114}$$

In this solution the particles have low velocities near the axis, but large number densities. Hence such particles do not contribute greatly to the current, but do have a significant effect in determining the potential difference between the beam centre and edge.

2. *Ultra-Relativistic Particles*   In this situation, $\gamma_v(a) \gg 1$, so that for all the charges except those very close to the beam axis, $\beta_v \approx 1$, and almost totally independent of radius. This has the consequence

$$\nu_B \approx \gamma_v(a) \tag{8.115}$$

and so the current in the beam is approximately the Alfvén current:

$$I \approx I_A \tag{8.116}$$

8.7 HYDROMAGNETIC SHOCKS

A shock is the transition between two different uniform gas states, though in practice, the gas behind the shock is not uniform. Hydromagnetic shocks have been extensively reviewed in [11, 83], and summarised for example in [15, 98, 73]. This section is concerned with plane shocks moving in the direction normal to the plane, in which the hydromagnetic equations (see Section 8.3) are valid on either side of the shock (but not actually inside the shock itself). Region 1 is the undisturbed, static region ahead of the shock, and Region 2 is the shocked region behind the shock; subscripts 1 and 2 will identify quantities ahead and behind the shock, respectively.

The notation

$$[\![Q]\!] = Q_2 - Q_1 \tag{8.117}$$

denotes the jump in the value of a quantity Q on either side of the shock.

In order to simplify the algebra, the equations are formulated in the rest frame of the shock, with a unit vector $\hat{\mathbf{n}}$ orthogonal to the shock plane, pointing in the direction of the shock propagation. Hence in all the following analysis,

$$\boldsymbol{u}_1 = \hat{\mathbf{x}} \, u_1 \tag{8.118}$$

so that the shock is propagating along the x-axis, and the unshocked material is assumed to be at rest. The transverse direction will be taken to be the y-direction. Then the hydromagnetic jump conditions are

$$[\![\hat{\mathbf{n}} \cdot \boldsymbol{B}]\!] = 0 \tag{8.119}$$

$$[\![\hat{\mathbf{n}} \times (\boldsymbol{u} \times \boldsymbol{B})]\!] = 0 \tag{8.120}$$

$$[\![\rho \hat{\mathbf{n}} \cdot \boldsymbol{u}]\!] = 0 \tag{8.121}$$

$$[\![\rho(\hat{\mathbf{n}} \cdot \boldsymbol{u})\boldsymbol{u} + \hat{\mathbf{n}}(p + B^2/(2\mu_0)) - (\hat{\mathbf{n}} \cdot \boldsymbol{B})\boldsymbol{B}/\mu_0]\!] = 0 \quad \text{(SI)} \tag{8.122}$$

$$[\![\rho(\hat{\mathbf{n}} \cdot \boldsymbol{u})\boldsymbol{u} + \hat{\mathbf{n}}(p + B^2/(8\pi)) - (\hat{\mathbf{n}} \cdot \boldsymbol{B})\boldsymbol{B}/(4\pi)]\!] = 0 \quad \text{(cgs)} \tag{8.123}$$

$$[\![(\hat{\mathbf{n}} \cdot \boldsymbol{u})\left(\rho e + \tfrac{1}{2}\rho u^2 + p + B^2/\mu_0\right) - (\hat{\mathbf{n}} \cdot \boldsymbol{B})(\boldsymbol{B} \cdot \boldsymbol{u})/\mu_0]\!] = 0 \quad \text{(SI)} \tag{8.124}$$

$$[\![(\hat{\mathbf{n}} \cdot \boldsymbol{u})\left(\rho e + \tfrac{1}{2}\rho u^2 + p + B^2/(4\pi)\right) - (\hat{\mathbf{n}} \cdot \boldsymbol{B})(\boldsymbol{B} \cdot \boldsymbol{u})/(4\pi)]\!] = 0 \quad \text{(cgs)} \tag{8.125}$$

where ρ is the mass density, $\boldsymbol{u}$ is the hydromagnetic fluid velocity in the shock rest frame, $\boldsymbol{B}$ is the magnetic field, p is the scalar hydromagnetic pressure, and e is the internal energy, defined by

$$e = \frac{p}{(\gamma - 1)\rho} \tag{8.126}$$

where $1 < \gamma < 2$ is the polytropic (or adiabatic) index. Note in particular that (8.119) demands continuity of the normal magnetic field component in all shocks. Expressions (8.119)-(8.125) are the generalisations of the Rankine-Hugoniot relations for hydrodynamical shocks. An alternative form of (8.124) is

$$[\![e]\!] + \tfrac{1}{2}(p_1 + p_2)\,[\![1/\rho]\!] + [\![\boldsymbol{B}]\!]^2\,[\![1/\rho]\!]\,/(4\mu_0) = 0 \quad \text{(SI)} \tag{8.127}$$

$$[\![e]\!] + \tfrac{1}{2}(p_1 + p_2)\,[\![1/\rho]\!] + [\![\boldsymbol{B}]\!]^2\,[\![1/\rho]\!]\,/(16\pi) = 0 \quad \text{(cgs)} \tag{8.128}$$

Note, from (8.119), that the normal component of the magnetic field is continuous, and from (8.121), that the mass flux through the shock is also continuous.

8.7.1 Further Notation

Introduce the ratios $\mathcal{R}_p$ and $\mathcal{R}_\rho$ which are defined as

$$\mathcal{R}_p = \frac{p_2}{p_1} \tag{8.129}$$

$$\mathcal{R}_\rho = \frac{\rho_2}{\rho_1} \tag{8.130}$$

and define the angle θ_j, $j = 1, 2$ between the magnetic field direction and the shock propagation direction as

$$\sin\theta_j = \frac{B_{j,y}}{B_j}, \quad j = 1, 2 \tag{8.131}$$

so that $\theta_1 = 0$ refers to a magnetic field which is parallel to $\hat{\mathbf{n}}$. Note that a shock is *compressive* if $\mathcal{R}_\rho > 1$, *non-compressive* if $\mathcal{R}_\rho = 1$ and *expansive* if $\mathcal{R}_\rho < 1$.

The ratio $\sigma_j, j = 1, 2$ of the square of the sound to Alfvén speeds on either side of the shock front is given by

$$\sigma_j = \frac{\mu_0 \gamma p_j}{B_j^2} \quad j = 1, 2 \quad \text{(SI)} \tag{8.132}$$

$$= \frac{4\pi \gamma p_j}{B_j^2} \quad \quad \text{(cgs)} \tag{8.133}$$

The reduced Alfvén speed $\alpha_j, j = 1, 2$ in the shock propagation direction is defined as

$$\alpha_j = \left(\frac{B_n^2}{\mu_0 \rho_j}\right)^{1/2}, \quad j = 1, 2, \quad B_n = \hat{\mathbf{n}} \cdot \mathbf{B} \qquad \text{(SI)} \qquad (8.134)$$

$$= \left(\frac{B_n^2}{4\pi \rho_j}\right)^{1/2} \qquad \qquad \text{(cgs)} \qquad (8.135)$$

The shock strength parameter h is a measure of the discontinuity in the transverse magnetic field, and is defined by

$$h = \frac{[\![B_y]\!]}{B_1} \qquad (8.136)$$

where B_1 is the magnitude of the *total* magnetic field in the unshocked region.

Finally, the mass flux through the shock is denoted $\mathcal{F}_\rho$, and defined by

$$\mathcal{F}_\rho = |\rho u_n|, \qquad u_n = \hat{\mathbf{n}} \cdot \mathbf{u} \qquad (8.137)$$

Note that the mass flux is directed oppositely to $\hat{\mathbf{n}}$.

8.7.2 Shock Classification

Following [11] we classify shocks as follows:

Contact Discontinuity Here there is no mass flow through the shock front. There are two subdivisions:

$\hat{\mathbf{n}} \cdot \mathbf{B} = 0$: shear flow and/or magnetic contact discontinuity are possible.

$\hat{\mathbf{n}} \cdot \mathbf{B} \neq 0$: neither shear flow nor magnetic contact discontinuity are possible.

In other words, a magnetic discontinuity or a shear flow discontinuity are only permitted if the magnetic field normal to the surface of discontinuity vanishes.

Non-compressive Shocks Also known as the *Transverse Alfvén Shock* and the *Intermediate shock*, this non-trivial solution for the case $[\![\rho]\!] = 0$ yields

$$[\![\hat{\mathbf{n}} \cdot \mathbf{u}]\!] = 0 \qquad (8.138)$$

$$[\![p]\!] = 0 \qquad (8.139)$$

$$[\![\mathbf{u}_{tr}]\!] = \text{sgn}(\hat{\mathbf{n}} \cdot \mathbf{B}) \, [\![\mathbf{B}_{tr}]\!] \qquad (8.140)$$

where $\mathbf{u}_{tr} = \mathbf{u} - \hat{\mathbf{n}}(\hat{\mathbf{n}} \cdot \mathbf{u})$, with $\mathbf{B}_{tr}$ defined similarly, and where the signum function sgn is defined as follows:

$$\text{sgn}(x) = \begin{cases} -1 & x < 0, \\ 0 & x = 0, \\ 1 & x > 0. \end{cases} \qquad (8.141)$$

In fact this solution corresponds to the passage of a large, finite amplitude Alfvén wave, and so is not really a true shock at all.

Fast Magnetic Shocks $B_{2,y} > B_{1,y} > 0$, $\hat{\mathbf{n}} \cdot \mathbf{B}_1 > 0$.   These shocks are compressive, and can be subdivided into 2 further classes:

$$\text{Type 1:} \qquad \sigma_1 \geq 1 - \frac{\gamma}{\gamma - 1} \sin^2 \theta_1 \quad \text{FM1} \qquad\qquad (8.142)$$

$$\text{Type 2:} \qquad \sigma_1 < 1 - \frac{\gamma}{\gamma - 1} \sin^2 \theta_1 \quad \text{FM2} \qquad\qquad (8.143)$$

Slow Magnetic Shocks $B_{2,y} < B_{1,y}$, $B_{1,y} > 0$, $\hat{\mathbf{n}} \cdot \mathbf{B} > 0$. These shocks are compressive, and can be subdivided into 2 further classes:

$$\text{Type 1:} \qquad \sigma_1 \geq 1 - \gamma \sin^2 \theta_1 \quad \text{SM1} \qquad\qquad (8.144)$$

$$\text{Type 2:} \qquad \sigma_1 < 1 - \gamma \sin^2 \theta_1 \quad \text{SM2} \qquad\qquad (8.145)$$

The limits $\theta_1 \to 0$ and $\theta_1 \to \pi/2$ are special cases:

Parallel Shocks $\theta_1 \to 0$, $B_{1,y} \to 0$, $\hat{\mathbf{n}} \cdot \mathbf{B} > 0$

FM1	$\sigma_1 \geq 1$	Fast pure gas shock	(8.146)
FM2	$\sigma_1 < 1$	Switch-on shock, fast gas shock	(8.147)
SM1	$\sigma_1 \geq 1$	Continuous transition	(8.148)
SM2	$\sigma_1 < 1$	Switch-on shock, slow gas shock	(8.149)

Perpendicular Shocks $\theta_1 \to \pi/2$, $\hat{\mathbf{n}} \cdot \mathbf{B} \to 0$ Note that only Type 1 shocks persist here, since σ has to be positive, and therefore the Type 2 constraint is not physically meaningful in this limit.

FM1	$\sigma_1 \geq 0$	Perpendicular shock	(8.150)
SM1	$\sigma_1 \geq 0$	Contact discontinuity	(8.151)

8.7.3   Shock Propagation Parallel to B_1

There are two special cases for parallel propagation:

8.7.3.1 Fast Pure Gas Shock (FM1)
The magnetic field has no effect, $\sigma_1 \geq 1$, and the jump conditions yield (eventually)

$$[\![u_y]\!] = 0 \tag{8.152}$$

$$[\![\boldsymbol{B}]\!] = 0 \tag{8.153}$$

$$[\![\rho u_n]\!] = 0 \tag{8.154}$$

$$\mathcal{R}_p = 1 + 2\gamma \frac{\mathcal{R}_\rho - 1}{1 + \gamma - (\gamma - 1)\mathcal{R}_\rho} \tag{8.155}$$

$$\mathcal{F}_\rho^2 = 2\gamma \frac{p_1}{2 - (\gamma - 1)(\mathcal{R}_\rho - 1)} \tag{8.156}$$

where $\sigma_1 \geq 1$ and

$$1 < \mathcal{R}_\rho \leq \frac{\gamma + 1}{\gamma - 1} \tag{8.157}$$

Note that this is termed a *fast* shock because $u_{n1} \geq \alpha_1$ for all permitted values of $\mathcal{R}_\rho$ as defined in (8.157), given that $\sigma_1 \geq 1$.

8.7.3.2 Switch-On Shock (FM2)
The solution here is a combination of a fast gas shock and a magnetic shock, such that $\sigma_1 < 1$. The jump conditions yield (eventually)

$$[\![B_y]\!]^2 = 2B_1^2 (\mathcal{R}_\rho - 1) \left[1 - \sigma_1 - \tfrac{1}{2}(\gamma - 1)(\mathcal{R}_\rho - 1) \right] \tag{8.158}$$

$$[\![u_n]\!] = \alpha_1 \left(\mathcal{R}_\rho^{1/2} - \mathcal{R}_\rho^{-1/2} \right) \tag{8.159}$$

$$[\![u_y]\!] = \alpha_2 \frac{B_{2y}}{B_n} = \alpha_1 \mathcal{R}_\rho^{-\frac{1}{2}} \frac{B_{1y}}{B_n} \tag{8.160}$$

$$\mathcal{F}_\rho = \rho_1 \alpha_1 \tag{8.161}$$

$$\mathcal{R}_p = 1 + \gamma (\mathcal{R}_\rho - 1) \left[1 + \frac{\gamma - 1}{2\sigma_1} (\mathcal{R}_\rho - 1) \right] \tag{8.162}$$

where this solution is valid in the range

$$1 + \frac{2}{\gamma - 1} (1 - \sigma_1) \leq \mathcal{R}_\rho < \frac{\gamma + 1}{\gamma - 1} \tag{8.163}$$

or equivalently,

$$1 + 2\frac{\gamma}{\gamma - 1} \left(\frac{1 - \sigma_1}{\sigma_1} \right) \leq \mathcal{R}_p < \infty \tag{8.164}$$

8.7.3.3 Switch-On Shock (SM2)   This shock is a combination of a magnetic shock and a slow gas shock. Again, $\sigma_1 < 1$, with the gas contribution being as given in Section 8.7.3.1, but with the restriction

$$1 < \mathcal{R}_\rho \leq 1 + 2\frac{1 - \sigma_1}{\gamma - 1} \tag{8.165}$$

The magnetic contribution is as given in Section 8.7.3.2, with the only difference being that the 'switched on' tangential magnetic component is in the opposite direction to that in Section 8.7.3.2.

8.7.4 Shock Propagation Perpendicular to B_1

Here the magnetic field is perpendicular to the direction of propagation of the shock. Solutions here are necessarily of Type 1 only. Recall from (8.119) that there can be no magnetic field component normal to the shock plane.

8.7.4.1 Perpendicular Shock (FM1)   The jump conditions yield

$$\mathcal{R}_\rho = \frac{B_{2y}}{B_{1y}} \tag{8.166}$$

$$[\![u_y]\!] = 0 \tag{8.167}$$

$$[\![u_n]\!] = c_{a1}\left[\frac{1 + \sigma_1 + \frac{1}{2}(2 - \gamma)\mathcal{R}_\rho}{1 - \frac{1}{2}(\gamma - 1)(\mathcal{R}_\rho - 1)}\right]^{1/2}\left(\mathcal{R}_\rho^{1/2} - \mathcal{R}_\rho^{-1/2}\right) \tag{8.168}$$

$$u_{n2} = \frac{u_{n1}}{\mathcal{R}_\rho} = \frac{c_{a1}}{\mathcal{R}_\rho^{1/2}}\left[\frac{1 + \sigma_1 + \frac{1}{2}(2 - \gamma)\mathcal{R}_\rho}{1 - \frac{1}{2}(\gamma - 1)(\mathcal{R}_\rho - 1)}\right]^{1/2} \tag{8.169}$$

$$\mathcal{R}_p = 1 + \gamma\mathcal{R}_\rho\frac{1 + (\gamma - 1)\mathcal{R}_\rho^2/(4\sigma_1)}{1 - \frac{1}{2}(\gamma - 1)(\mathcal{R}_\rho - 1)} \tag{8.170}$$

where the validity of these results demands

$$1 < \mathcal{R}_\rho \leq \frac{\gamma + 1}{\gamma - 1} \tag{8.171}$$

and moreover

$$u_{n1} > \left(c_{a1}^2 + c_{th1}^2\right)^{1/2} \tag{8.172}$$

$$u_{n2} < \left(c_{a2}^2 + c_{th2}^2\right)^{1/2} \tag{8.173}$$

The symbols $c_{aj}, c_{thj}, \quad j = 1, 2$ denote the Alfvén and thermal speeds ahead of, and behind, the shock front, respectively.

8.7.4.2 Contact Discontinuity
This $\theta_1 = \pi/2$ limit of the SM1 shock is a contact discontinuity, in which

$$\mathcal{F}_\rho = 0 \tag{8.174}$$

$$\left[\!\left[p + \frac{B^2}{2\mu_0} \right]\!\right] = 0 \quad \text{(SI)} \tag{8.175}$$

$$\left[\!\left[p + \frac{B^2}{8\pi} \right]\!\right] = 0 \quad \text{(cgs)} \tag{8.176}$$

$$\frac{\rho_2}{\rho_1} = 1 + \frac{(3 - B_{2y}/B_1)(B_{2y}/B_1 - 1)}{2\sigma_1 + (\gamma - 1)(B_{2y}/B_1 - 1)} \tag{8.177}$$

$$[\![u_y]\!] = -\alpha_1 (2 - B_{2y}/B_1 - 1)(B_{2y}/B_1 - 1) \times$$

$$\left[\frac{2(1 + \sigma_1) + (\gamma - 2)(B_{2y}/B_1 - 1)}{2\sigma_1 + (\gamma - 1)(B_{2y}/B_1 - 1) - (B_{2y}/B_1 - 1)^2} \right]^{1/2} \tag{8.178}$$

8.7.5 General Case: Fast Magnetic Shocks

$$h > 0, \quad 0 < \theta_1 < \pi/2$$

The shock is described by

$$\mathcal{R}_\rho = 1 + h \frac{-\frac{1}{2}\gamma h \sin\theta_1 - (1 - \sigma_1) \pm R^{1/2}}{2\sigma_1 \sin\theta_1 - (\gamma - 1)h} \tag{8.179}$$

$$R = h^2 (\tfrac{1}{4}\gamma^2 \sin^2\theta_1 + 1 - \gamma) + h(2 - \gamma)(1 + \sigma_1)\sin\theta_1$$
$$+ 4\sigma_1 \sin^2\theta_1 + (1 - \sigma_1)^2 \tag{8.180}$$

$$\mathcal{R}_p = 1 + \frac{\gamma h}{\sigma_1} \left[-\tfrac{1}{2}h + \frac{(\mathcal{R}_\rho - 1)/h - \sin\theta_1}{1 - \sin\theta_1(\mathcal{R}_\rho - 1)/h} \right] \tag{8.181}$$

$$[\![u_y]\!] = \frac{h\alpha_2}{\cos\theta_1} \left[1 - \frac{\mathcal{R}_\rho - 1}{h}\sin\theta_1 \right]^{1/2} \tag{8.182}$$

$$[\![u_n]\!] = \alpha_1 \frac{\mathcal{R}_\rho - 1}{[1 - (\mathcal{R}_\rho - 1)/h \sin\theta_1]^{1/2}} \tag{8.183}$$

Class 1 shocks (FM1) depend only on the '+' sign in (8.179), and satisfy

$$\sigma_1 \geq 1 - \frac{\gamma}{\gamma - 1}\sin^2\theta_1 \tag{8.184}$$

$$0 < h < 2\frac{\sin\theta_1}{\gamma - 1} \tag{8.185}$$

Class 2 shocks (FM2) have both branches of (8.179) present, and satisfy

$$\sigma_1 < -\frac{\gamma}{\gamma - 1} \sin^2 \theta_1 \qquad (8.186)$$

with the '+' branch yielding part of the total solution when h satisfies

$$0 < h < h^* \qquad (8.187)$$

where

$$h^* = \frac{\sin \theta_1 (2 - \gamma)(1 + \sigma_1) + 2 \cos \theta_1 \left[(\gamma - 1)(1 - \sigma_1)^2 + \sigma_1 \gamma^2 \sin^2 \theta_1\right]^{1/2}}{2(\gamma - 1) - \frac{1}{2}\gamma^2 \sin^2 \theta_1}$$

$$(8.188)$$

and the remaining part, from the '−' branch, contributing subject to the restriction

$$2\frac{\sin \theta_1}{\gamma - 1} < h < h^* \qquad (8.189)$$

8.7.6 General Case: Slow Magnetic Shocks

$$h < 0, \quad 0 < \theta_1 < \pi/2$$

The shock here is described by the same relations as for the Fast Magnetic Shock, (8.179-8.183) with $-h$ substituted for h.

Note that in a slow magnetic shock, the magnitude of the transverse magnetic field behind the shock is always less than or equal to that in the unshocked material ahead. For this reason, these shocks are sometimes referred to as 'switch-off' shocks.

Class 1 shocks (SM1) are restricted to the '+' roots only, with the additional restriction that

$$\sigma_1 \geq 1 - \gamma \sin^2 \theta_1 \qquad (8.190)$$

$$0 > h > -2 \sin \theta_1 \qquad (8.191)$$

Class 2 shocks (SM2)

$$\sigma_1 < 1 - \gamma \sin^2 \theta_1 \qquad (8.192)$$

The solution here can depend on both branches, with the '+' root admissable provided

$$0 > h \geq h^\dagger \qquad (8.193)$$

where

$$h^\dagger = -\frac{-\sin\theta_1(2-\gamma)(1+\sigma_1) + 2\cos\theta_1\left[(\gamma-1)(1-\sigma_1)^2 + \sigma_1\gamma^2\sin^2\theta_1\right]}{2(\gamma-1) - \frac{1}{2}\gamma^2\sin^2\theta_1}$$

(8.194)

and the '−' branch permitted if

$$-2\sin\theta_1 \geq h \geq h^\dagger$$

(8.195)

8.7.7   Further Reading

These results on MHD shocks can be generalised to the case where the ambient hydromagnetic flow is not aligned with the shock propagation direction; see [62] for details.

8.8   ION-ACOUSTIC SHOCK

A two-fluid plasma contains hot electrons, at a temperature T_e, and singly-charged cold ions, such that the electrons satisfy Boltzmann statistics,

$$n_e = n_0 \exp\left(\frac{e\phi}{k_B T_e}\right)$$

(8.196)

where n_0 is the equilibrium ion number density (so that the plasma is electrically neutral overall) and ϕ is the electric potential.

Assuming a 1-D treatment only, the ions satisfy the equations

$$\frac{\partial n_i}{\partial t} + \frac{\partial n_i u_i}{\partial x} = 0$$

(8.197)

$$\frac{\partial u_i}{\partial t} + u_i\frac{\partial u_i}{\partial x} = -\frac{e}{m_i}\frac{\partial \phi}{\partial x}$$

(8.198)

$$\frac{\partial^2 \phi}{\partial x^2} = -\frac{e}{\epsilon_0}\left[n_i - n_0\exp\left(\frac{e\phi}{k_B T_e}\right)\right] \quad \text{(SI)}$$

(8.199)

$$= -4\pi e\left[n_i - n_0\exp\left(\frac{e\phi}{k_B T_e}\right)\right] \quad \text{(cgs)}$$

(8.200)

in which n_i is the ion number density, m_i the ion mass, and u_i the ion speed in the x-direction.

In the steady state, search for a travelling waveform solution, so that all ion quantities can be written as functions of $\xi = x - Ut$, where U will be the phase speed of the waveform. In this way,

$$u_i = U \pm \left(U^2 - \frac{2e}{m_i}\phi\right)^{1/2} \tag{8.201}$$

$$n_i = n_0 \left(1 - \frac{2e\phi}{k_B T_e}\right)^{-1/2} \tag{8.202}$$

Putting $y = e\phi/(k_B T_e)$ allows (8.199) to be written as

$$\frac{d^2 y}{d\xi^2} = -\frac{1}{\lambda_D^2}\left[\left(1 - \frac{2y}{\mathcal{M}^2}\right)^{-1/2} - \exp(y)\right] \tag{8.203}$$

where

$$\lambda_D^2 = \frac{\epsilon_0 k_B T_e}{n_0 e^2} \quad \text{(SI)} \tag{8.204}$$

$$= \frac{k_B T_e}{4\pi n_0 e^2} \quad \text{(cgs)} \tag{8.205}$$

$$\mathcal{M} = \frac{U}{k_B T_e / m_i} \tag{8.206}$$

Demanding $dy/d\xi \to 0$ as $y \to 0$ allows (8.203) to be written in the form

$$\frac{1}{2}\left(\frac{dy}{d\xi}\right)^2 = \frac{1}{\lambda_D^2}\left[\mathcal{M}^2\left(1 - \frac{2y}{\mathcal{M}^2}\right)^{1/2} + \exp(y) - \mathcal{M}^2 - 1\right] \tag{8.207}$$

Bounded and localised solutions of (8.207) require

$$1 < \mathcal{M}^2 < 2.56 \tag{8.208}$$

corresponding to that range of $\mathcal{M}$ over which the right-hand side of (8.207) has two roots.

If y is small, allowing the right-hand side of (8.207) or (8.203) to be expanded to order y^2 results in a non-linear equation that has a solitary-wave like solution of the form

$$y = y_0 \cosh^{-2}(\kappa\xi) \tag{8.209}$$

$$y_0 = 3\frac{\mathcal{M}^2(\mathcal{M}^2 - 1)}{3 - \mathcal{M}^4} \tag{8.210}$$

$$\kappa = \frac{(\mathcal{M}^2 - 1)^{1/2}}{2\lambda_D \mathcal{M}} \tag{8.211}$$

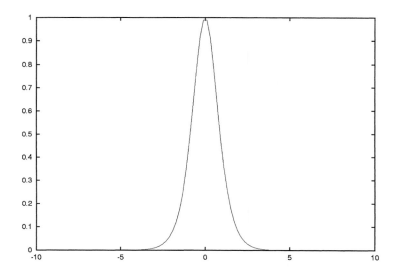

Fig. 8.1   The small-amplitude ion-acoustic shock solution $y = y_0 \cosh^{-2}(\kappa\xi)$.

as shown in Figure 8.1. Note that the width Δ of the shock can be approximated as

$$\Delta \approx \lambda_D \frac{\mathcal{M}}{\sqrt{\mathcal{M}^2 - 1}} \tag{8.212}$$

The full solution to (8.207) has to be obtained numerically, and can display several static peaks and troughs in the rest-frame of the shock.

Note that the ion acoustic shock is intimately related to the formation of plasma sheaths (see (3.5) in Section 3.2.1). In fact the ion-acoustic shock can be considered to be a sheath travelling through the plasma. However, the boundary conditions for a static sheath in a plasma bounded by electrodes are different from those discussed above.

9

Equilibria and Instabilities

9.1 NOTATION

Symbol	Meaning	Ref
a	plasma column radius	
$\boldsymbol{A}$	magnetic vector potential	
$\boldsymbol{B}$	magnetic flux density	
c	speed of light in vacuo	
c_a	Alfvén speed for the plasma	(2.24)
E_D	Dreicer electric field	(9.111)
$\boldsymbol{E}$	electric field	
$\boldsymbol{g}$	acceleration due to gravity	
I	current	
I_0	modified Bessel function of order 0	
I_n	modified Bessel function of order n	
$\boldsymbol{J}$	current density	
J_m	Bessel function of 1st kind, order m	
$\boldsymbol{k}$	wave-vector	
K_m	modified Bessel function, order m	
m_s	mass of particle of species s	
p	gas pressure	
q	safety factor	(9.64)
s	label defining species:i (ion), e (electron), n (neutral)	
T_s	temperature of gas of species s	
$\boldsymbol{u}$	fluid velocity	
γ	adiabatic index	
γ_g	growth rate	
η	fluid plasma resistivity	
η_v	plasma viscosity	
λ_D	Debye length	(2.17)
ν_B	Budker parameter	(8.87)
ρ	mass density of single-fluid plasma	(7.80)
σ_j	square of sound over Alfvén speed on either side of shock	(8.132)
τ_A	Alfvén transit time	(2.13)
τ_R	resistive diffusion time	(2.15)
ϕ_i	ionization potential	
ω	wave frequency	
ω_{cs}	circular cyclotron frequency of species s	(2.7)
ω_p	circular plasma frequency	(2.6)
$\boldsymbol{\omega}$	fluid vorticity	(8.6)

9.2 GENERAL CONSIDERATIONS

An equilibrium state is one which is not evolving in time, that is, no component of the system has an explicit time dependence. Equilibria can be stationary (no bulk motion) or dynamic (steady bulk motion); they can also be stable or unstable. A stable equilibrium is a state which if perturbed produces restoring forces which act to reverse the perturbation and re-establish the original equilibrium. Unstable equilibria do not possess the requisite restoring forces, and exhibit instability to perturbation. An instability can be defined as an unbounded growth away from an equilibrium configuration; if a quantity becomes unbounded in finite time, this is referred to as an explosive instability. A key to the common terminology follows:

type of instability	description
absolute	unbounded growth at all spatial points simultaneously
configuration-space	evolution of macroscopic quantities away from thermal equilibrium
convective	unbounded growth evolves as a disturbance propagates
electromagnetic	unbounded growth associated with accumulation of current density
electrostatic	unbounded growth associated with unconstrained charge accumulation
parametric	instability provoked by the application of an external periodic stimulus
velocity-space	evolution of kinetic distribution function away from Maxwellian equilibrium

More detailed classification systems for plasma instabilities are described in [21]

9.3 FLUID EQUILIBRIA

9.3.1 Ideal MHD

The classical stationary ideal MHD equilibrium is given by

$$\boldsymbol{u_0} = 0 \qquad\qquad\qquad (9.1)$$

$$\nabla p_0 = \boldsymbol{J}_0 \times \boldsymbol{B}_0 \qquad \text{(SI)} \qquad (9.2)$$

$$\nabla p_0 = \boldsymbol{J}_0 \times \boldsymbol{B}_0/c \qquad \text{(cgs)} \qquad (9.3)$$

where subscript 0 denotes an equilibrium quantity. Immediate consequences are

$$\boldsymbol{B}_0 \cdot \nabla p_0 = 0 \tag{9.4}$$

$$\boldsymbol{J}_0 \cdot \nabla p_0 = 0 \tag{9.5}$$

showing that the pressure is constant along lines of magnetic field, and also along lines of current density.

9.3.1.1 *Uniform $\boldsymbol{B}_0$*   Where the equilibrium magnetic field is spatially uniform, then

$$\boldsymbol{J}_0 = 0 \tag{9.6}$$

$$p_0 = \text{constant} \tag{9.7}$$

9.3.1.2 *General Case*   In general, (9.2) can be written in the form

$$\nabla \left(p_0 + \frac{B_0^2}{2\mu_0} \right) = \frac{(\boldsymbol{B}_0 \cdot \nabla)\boldsymbol{B}_0}{2\mu_0} \quad \text{(SI)} \tag{9.8}$$

$$\nabla \left(p_0 + \frac{B_0^2}{8\pi} \right) = \frac{(\boldsymbol{B}_0 \cdot \nabla)\boldsymbol{B}_0}{8\pi} \quad \text{(cgs)} \tag{9.9}$$

If the magnetic field is unidirectional in Cartesian co-ordinates, or axial in cylindrical co-ordinates, then (9.8) reduces to

$$p_0 + \frac{B_0^2}{2\mu_0} = \text{constant (SI)} \tag{9.10}$$

$$p_0 + \frac{B_0^2}{8\pi} = \text{constant (cgs)} \tag{9.11}$$

9.3.1.3 *Force-Free Equilibrium*   If $\boldsymbol{J}_0 \neq 0$ and is parallel to $\boldsymbol{B}_0$, then a force-free equilibrium exists, in which

$$\boldsymbol{J}_0 \times \boldsymbol{B}_0 = 0 \tag{9.12}$$

$$\nabla \times \boldsymbol{B}_0 = \alpha \boldsymbol{B}_0 \tag{9.13}$$

$$p_0 = \text{constant} \tag{9.14}$$

where α characterises the equilibrium configuration. If α is a constant, then the magnetic fields determined by (9.13) correspond to the lowest magnetic energy states which a closed system may attain [97]. Moreover, constant α force-free fields are a subset of a wider class of equilibria satisfying

$$\nabla \times (\nabla \times \boldsymbol{B}) = \alpha^2 \boldsymbol{B} \tag{9.15}$$

which have the maximum magnetic energy density for a given current density, or equivalently, have minimum magnetic dissipation for a given magnetic energy [22], neglecting surface currents.

9.3.1.4 Taylor Equilibria Note that the concept of force-free equilibria as described by (9.13) may extended to resistive plasmas [91], provided that an extra constraint is satisfied, namely the total magnetic helicity $\boldsymbol{K}_0$ of the plasma is invariant [97]:

$$\boldsymbol{K}_0 = \int_{V_0} \boldsymbol{A} \cdot \boldsymbol{B} \mathrm{d}\tau = \text{ constant} \qquad (9.16)$$

where $\boldsymbol{A}$ is the magnetic vector potential, and V_0 is the total plasma volume. Under the constraint (9.16) a resistive plasma surrounded by a perfectly conducting toroidal shell will relax to a minimum energy state characterised by (9.13), where α is now directly related to the total current, the toroidal magnetic field and the plasma minor radius. (See [91] for detailed discussion of such equilibria.)

9.3.2 Cylindrical Equilibria

9.3.2.1 Bennett Relation An ideal MHD cylindrical plasma of radius a bounded by vacuum and carrying a total current I satisfies (9.2) in equilibrium. If the magnetic field is the self-field arising from the plasma current, then the Bennett relation states [52, 54]

$$I^2 = \frac{8\pi}{\mu_0} N_e k_B (T_e + T_i) \quad \text{(SI)} \qquad (9.17)$$

$$= \frac{16\pi}{\mu_0} N_e k_B T \qquad (9.18)$$

$$I^2 = 2c^2 N_e k_B (T_e + T_i) \quad \text{(cgs)} \qquad (9.19)$$

$$= 4c^2 N_e k_B T \qquad (9.20)$$

where N_e is the total number of electrons per unit axial length of the plasma cylinder, T_e, T_i are the electron and ion temperatures, respectively, and T is the plasma temperature for the equal temperature plasma case. Note that the temperature of each species is assumed to be spatially constant, and that the ions are singly charged.

The Bennett relation (9.17) can be expressed in the equivalent form [54]

$$\frac{< v_\phi^2 >}{v_z^2} = \frac{\nu_B}{\gamma_v} \qquad (9.21)$$

$$= \frac{a^2 \omega_p^2}{4c^2} \qquad (9.22)$$

where $< \cdots >$ denotes the average over beam radius, v_ϕ is the azimuthal speed, v_z is the axial speed, ν_B is the Budker parameter (8.87) and γ_v is the relativistic factor,

$$\gamma_v = \left[1 - \frac{v_\phi^2 + v_z^2}{c^2} \right]^{-\frac{1}{2}} \tag{9.23}$$

Relation (9.22) holds only if the beam is spatially homogeneous, so that (8.89) can be used.

9.3.2.2 Plasma Column Resonances

An unmagnetised electrically neutral plasma column of radius a, containing static ions and thermal electrons characterised by a scalar pressure p and associated temperature T, is surrounded by vacuum and can be driven to resonant oscillation by absorption of electromagnetic radiation at specific frequencies ω, given by [50]

$$\frac{n}{ka} \frac{J_n(ka)}{J_n'(ka)} = 2\frac{\omega^2}{\omega_p^2} - 1 \tag{9.24}$$

where J_n is the Bessel function of order n, J_n' is the derivative of J_n with respect to argument, and

$$ka = \frac{1}{\sqrt{3}} \left(\frac{a}{\lambda_D} \frac{\omega^2}{\omega_p^2} - 1 \right)^{\frac{1}{2}} \tag{9.25}$$

Note that (i)(9.24) corresponds to modes for which there is no surface charge density at the plasma-vacuum boundary, and therefore no radial current density at the boundary; (ii) 1-D adiabatic compression of the electrons is assumed, so that the polytropic index γ is given by $\gamma = 3$.

The main resonance occurs at $n = 0$, yielding $\omega = \omega_p/\sqrt{2}$. Higher order resonances occur at frequencies given by [50]

$$\omega^2 = \omega_p^2 \left(1 + 3\frac{\lambda_D^2}{a^2}x_n^2 \right) \tag{9.26}$$

where $x_1 \approx 5.3$, $x_2 \approx 8.5$. Further resonances can be calculated graphically from the solution of (9.24).

9.3.2.3 Surface Waves on a Plasma Cylinder

Let the unmagnetised cylindrical plasma of radius a be surrounded by a conducting cylinder of radius $b \geq a$.

unmagnetised cold plasma partially filling a conducting waveguide The dispersion relation for azimuthally symmetric E modes (that is, waves with an axial electric field only) is [50, 90, 92]

$$\left(1 - \frac{\omega_p^2}{\omega^2} \right) \frac{\kappa_0}{\kappa} \frac{I_0'(\kappa a)}{I_0(\kappa a)} = \frac{I_0'(\kappa_0 a)K_0(\kappa_0 b) - I_0(\kappa_0 b)K_0'(\kappa_0 a)}{I_0(\kappa_0 a)K_0(\kappa_0 b) - I_0(\kappa_0 b)K_0(\kappa_0 a)} \tag{9.27}$$

where

$$\kappa_0^2 = k^2 - \omega/c^2 \tag{9.28}$$

$$\kappa^2 = k^2 - (\omega^2 - \omega_p^2)/c^2 \tag{9.29}$$

and where the axial electric field is given by

$$E_z = \begin{cases} A\dfrac{I_0(\kappa r)}{I_0(\kappa a)}e^{i(kz-\omega t)} & 0 < r < a \\[2ex] A\dfrac{I_0(\kappa_0 r)K_0(\kappa_0 b) - I_0(\kappa_0 b)K_0(\kappa_0 r)}{I_0(\kappa_0 a)K_0(\kappa_0 b) - I_0(\kappa_0 b)K_0(\kappa_0 a)}e^{i(kz-\omega t)} & a < r < b \end{cases} \tag{9.30}$$

I_0 and K_0 are the modified Bessel functions of the first and second kind, respectively, and $'$ denotes the derivative with respect to argument. These waves propagate at less than the speed of light, and at frequencies below the waveguide cut-off.

isolated cylinder   In the limit of $b \gg a$, (9.27) is replaced by the approximate form

$$1 - \frac{\omega_p^2}{\omega^2} = \frac{\kappa}{\kappa_0}\frac{I_0(\kappa a)K_0'(\kappa_0 a)}{I_0'(\kappa a)K_0(\kappa_0 a)} \tag{9.31}$$

which is the dispersion relation for slow surface waves on a plasma column of radius a in a vacuum.

plasma filled cylinder, infinite magnetic field   Where the plasma fills the cylinder, so that $a = b$, and the axial magnetic field is so large that electrons are constrained to move parallel to the axis only, then only the E modes are affected by the plasma, with the axial electric field given by [50, 90, 92]

$$E_z = AJ_n(k_\perp r)e^{i(kz+m\theta-\omega t)} \tag{9.32}$$

with dispersion relation

$$k^2 a^2 = \frac{\omega^2 a^2}{c^2} - \frac{\alpha_{n\nu}^2}{1 - \omega_p^2/\omega^2} \tag{9.33}$$

where $\alpha_{n\nu}$ is the νth zero of the Bessel function J_n, and

$$k_\perp^2 = (\omega^2/c^2 - k^2)(1 - \omega_p^2/\omega^2) \tag{9.34}$$

satisfies the appropriate boundary condition $k_\perp a = \alpha_{n\nu}$.

9.4 FLUID INSTABILITIES

9.4.1 Firehose Instability

In an ideal MHD plasma with a magnetic field $\boldsymbol{B}_0 = \hat{\boldsymbol{z}} B_0$. Instead of a scalar pressure, assume a pressure tensor in the form

$$\mathbf{P} = \begin{bmatrix} p_\perp & 0 & 0 \\ 0 & p_\perp & 0 \\ 0 & 0 & p_\| \end{bmatrix} \tag{9.35}$$

where $p_\|$ is the pressure in the direction of the magnetic field, and $p_\perp$ is the pressure in the plane perpendicular to $\boldsymbol{B}$. Then two equations of state are required:

$$\frac{p_\perp^2 p_\|}{\rho} = \text{constant} \tag{9.36}$$

$$\frac{p_\perp}{\rho B} = \text{constant} \tag{9.37}$$

The dispersion relation for waves of frequency ω and wavevector $\boldsymbol{k}$ propagating at an angle θ to the magnetic field is

$$\omega^2 = \frac{k^2}{2\rho_0}(a \pm b) \tag{9.38}$$

where (in SI):

$$a = B_0^2/\mu_0 + p_{\perp 0} + 2p_{\|0}\cos^2\theta + p_{\perp 0}\sin^2\theta \tag{9.39}$$

$$b = \left\{ \left[B_0^2/\mu_0 + p_{\perp 0}(1 - \sin^2\theta) - 4p_{\|0}\cos^2\theta \right]^2 + 4p_{\|0}\sin^2\theta\cos^2\theta \right\}^{1/2} \tag{9.40}$$

In cgs units:

$$a = B_0^2/(4\pi) + p_{\perp 0} + 2p_{\|0}\cos^2\theta + p_{\perp 0}\sin^2\theta \tag{9.41}$$

$$b = \left\{ \left[B_0^2/(4\pi) + p_{\perp 0}(1 - \sin^2\theta) - 4p_{\|0}\cos^2\theta \right]^2 + 4p_{\|0}\sin^2\theta\cos^2\theta \right\}^{1/2} \tag{9.42}$$

For the particular case of propagation parallel to the magnetic field, (9.38) has the two solutions

$$\omega^2 = \frac{k^2}{\rho_0}\left(B_0^2/\mu_0 + p_{\perp 0} - p_{\|0}\right) \quad \text{(SI)} \tag{9.43}$$

$$= \frac{k^2}{\rho_0}\left(B_0^2/(4\pi) + p_\perp 0 - p_{\|0}\right) \quad \text{(cgs)} \tag{9.44}$$

$$\omega^2 = 3k^2\frac{p_{\|0}}{\rho_0} \tag{9.45}$$

Equation (9.43) shows that when $p_{\|0} > B_0^2/\mu_0 + p_{\perp 0}$, Alfvén waves are unstable. The other solution (9.45) corresponds to ion sound waves propagating along the magnetic field direction.

Physical mechanism: The pressure variation induced by the passing Alfvén wave becomes too great for the perpendicular pressure component of the plasma, and the restoring force falls short of that required to balance the magnetic variation.

9.4.2   Gravitational Instability

Also known as the *Rayleigh-Taylor* and *Kruskal-Schwarzschild* instability. A cold magnetised flowing plasma supported against gravity will become unstable in a manner analogous to Rayleigh-Taylor instability (see Figure 9.1). The equation governing the ion equilibrium is

$$m_i(\boldsymbol{u}_{i0} \cdot \nabla)\boldsymbol{u}_{i0} = q_i\boldsymbol{u}_{i0} \times \boldsymbol{B}_0 + m_i\boldsymbol{g} \qquad \text{(SI)} \qquad (9.46)$$

$$= q_i\boldsymbol{u}_{i0} \times \boldsymbol{B}_0/c + m_i\boldsymbol{g} \qquad \text{(cgs)} \qquad (9.47)$$

which yields, for constant gravity,

$$\boldsymbol{u}_{i0} = \frac{m_i}{q_i} \frac{\boldsymbol{g} \times \boldsymbol{B}_0}{B_0^2} \qquad \text{(SI)} \qquad (9.48)$$

$$= \frac{m_i}{cq_i} \frac{\boldsymbol{g} \times \boldsymbol{B}_0}{B_0^2} \qquad \text{(cgs)} \qquad (9.49)$$

Assuming that there is a density gradient ∇n_0 in the opposite direction to $\boldsymbol{g}$, and that the equilibrium flow is perpendicular to both $\boldsymbol{g}$ and $\boldsymbol{B}$, then the plasma is unstable to small perturbations going as $\exp[\mathrm{i}(kx - \omega t)]$ propagating along the flow direction, with a growth rate of

$$\gamma_g \approx \left[-g\frac{n_0'}{n_0}\right]^{1/2} \qquad (9.50)$$

where $'$ denotes the derivative.

This instability is also known as the *flute instability*, in cylindrical geometry, where the radial forces resulting from the magnetic curvature produce an equivalent gravity (see Figure 9.2). This is a simple example of the general class of such instabilities termed *interchange instabilities*, in which a magnetic field geometry which is concave to the plasma will be unstable to flute-type perturbations, whereas a convex magnetic field is stable.

Physical mechanism: Ion drift in the magnetised plasma due to the density gradient produces charge separation at the perturbed interface. At the critical wavelength, the resulting electric field creates an $\boldsymbol{E} \times \boldsymbol{B}$ drift which enhances the perturbation.

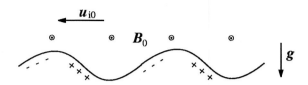

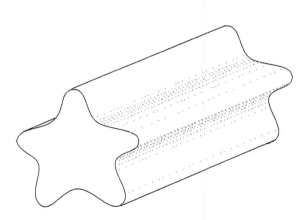

Fig. 9.1 Diagram showing the evolution of the charge separation under the gravitational instability. The resulting electric field can interact with the magnetic field to enhance the distortion, if the wave-length is right for correct phasing of the $E \times B$ drift.

Fig. 9.2 Diagram showing a flute instability of a cylindrical plasma.

9.4.3 Kelvin-Helmholtz Instability

The inviscid and incompressible flow of two horizontal infinite streams of ideal MHD plasma with a common horizontal plane interface perpendicular to the gravitational field g is unstable to interfacial perturbations going as $\exp[\mathrm{i}(kx - \omega t)]$, where x is the co-ordinate along the interface.

Let the density and speed of the lower plasma stream be ρ_1 and u_1, and those of the upper be ρ_2 and u_2, where u_1 and u_2 are directed parallel to the interface. Then the dispersion relation for interfacial waves is [98]

$$\omega = k\frac{\rho_1 u_1 + \rho_2 u_2}{\rho_1 + \rho_2} \pm k\left[c_a^2 - \frac{\omega_0^2}{k^2} - \frac{\rho_1 \rho_2}{(\rho_1 + \rho_2)^2}\right]^{1/2} \quad (9.51)$$

where ω_0 is the growth rate of the unmagnetised hydrodynamical instability, given by

$$\omega_0^2 = gk\frac{\rho_2 - \rho_1}{\rho_1 + \rho_2} \quad (9.52)$$

Physical mechanism: If the upper fluid is denser than the lower, any perturbation of the interface will result in a lower energy configuration, since the gain in gravitational potential energy resulting from lowering a dense fluid element is greater than the additional potential energy required to raise the lighter fluid. The magnetic field acts as an additional pressure contribution, which can stabilise the velocity discontinuity, if $|u_1 - u_2| < (\rho_1 + \rho_2)c_a/(\rho_1 \rho_2)^{1/2}$

9.4.4 Cylindrical Pinch Instabilities

A perfectly conducting isolated, self-confining and incompressible cylindrical MHD plasma of radius a bounded by vacuum carries an imposed, uniform, internal axial magnetic field B_{z0}, an imposed, uniform, external axial field B_{ze}, and an external, azimuthal, magnetic field arising from the surface current, so that $\boldsymbol{B} = \hat{\boldsymbol{\theta}}\,B_\theta(\rho) + \hat{\boldsymbol{z}}\,(B_{z0} + B_{ze})$ in cylindrical co-ordinates (ρ, θ, z). Perturbations of this equilibrium are assumed to take the form $f(\rho)\exp[\mathrm{i}(m\theta + kz - \omega t)]$, where $m = 0, 1, 2, \ldots$. The general dispersion relation for such perturbations is [19]

$$\frac{\omega^2}{k^2 c_a^2} = 1 - \left(\frac{B_{ze}}{B_{z0}} + \frac{m}{ka}\frac{B_\theta(a)}{B_{z0}}\right)^2 \frac{I'_m(ka)K_m(ka)}{I_m(ka)K'_m(ka)} - \frac{B_\theta^2(a)}{B_{z0}^2}\frac{1}{ka}\frac{I'_m(ka)}{I_m(ka)} \quad (9.53)$$

where I_m and K_m are the modified Bessel functions of the first and second kind, of order m, respectively, and

$$c_a^2 = B_{z0}^2/(\mu_0 \rho_0) \quad \text{(SI)} \quad (9.54)$$
$$= B_{z0}^2/(4\pi \rho_0) \quad \text{(cgs)} \quad (9.55)$$

Note that the azimuthal magnetic field component is confined to the plasma surface only, and so does not enter into the Alfvén speed (9.54).

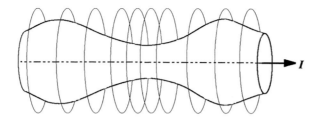

Fig. 9.3 Diagram showing the sausage instability of a cylindrical plasma. Note the increased magnetic field at the narrowest region.

9.4.4.1 Sausage Instability: m = 0 Axisymmetric perturbations have $m = 0$, and if in addition, $B_{ze} = 0$ satisfy the simplified dispersion relation

$$\frac{\omega^2}{k^2 c_a^2} = \left(1 - \frac{B_\theta^2(a)}{B_{z0}^2} \frac{1}{ka} \frac{I_0'(ka)}{I_0(ka)}\right) \tag{9.56}$$

Since $I_0'(x) < \frac{1}{2} x I_0(x)$, then instability occurs if

$$B_{z0}^2 < \tfrac{1}{2} B_\theta^2(a) \tag{9.57}$$

The generalisation of (9.57) to compressional plasmas is given in [15]:

$$\frac{\omega^2}{k^2 c_a^2} = 1 - \frac{B_\theta^2(a)}{B_{z0}^2} \frac{I_0'(Ka)}{KaI_0(Ka)} \frac{K^2}{k^2} \tag{9.58}$$

where

$$K = k\left[1 + \frac{\omega^4}{k^4 c_a^2 c_{th}^2 - \omega^2 k^2 (c_a^2 + c_{th}^2)}\right]^{1/2} \tag{9.59}$$

Instability requires

$$\frac{B_{z0}^2}{B_\theta^2(a)} \leq \frac{K^2}{k^2}\left(\frac{I_0'(Ka)}{KaI_0(Ka)}\right) < \tfrac{1}{2}\frac{K^2}{k^2} \tag{9.60}$$

Physical mechanism: The axisymmetric rippling of the cylindrical plasma produces regions where the plasma column is narrower than in equilibrium (Figure 9.3). In such regions, the azimuthal magnetic field is greater, since the current density is greater; hence the magnetic pressure here is enhanced, and tends to exacerbate the necking.

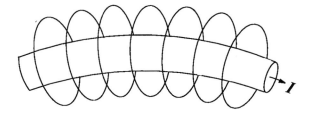

Fig. 9.4 Diagram showing the kink instability of a cylindrical plasma. Note the increased magnetic field where the plasma surface is convex with respect to the interior.

9.4.4.2   Kink Instability: $m \neq 0$ The general expression for an incompressible plasma is given in (9.53), which simplifies in the limit of long-wavelength perturbations, where $ka \ll 1$ [19]:

$$\frac{\omega^2}{k^2 c_a^2} \approx \left[1 + \left(\frac{B_{ze}}{B_{z0}} + \frac{m}{ka} \frac{B_\theta(a)}{B_{z0}} \right)^2 - \frac{m^2}{k^2 a^2} \frac{B_\theta(a)^2}{B_{z0}^2} \right] \qquad (9.61)$$

Instability occurs if ka is too small, that is, if the wavelength of the perturbation is too large. The Kruskal-Shafranov condition for stability is

$$\left| \frac{B_\theta}{B_z} \right| < \frac{2\pi a}{L} \qquad (9.62)$$

where L is maximum lengthscale in a finite plasma. The stability criterion (9.62) can also be written

$$q(a) > 1 \qquad (9.63)$$

where the safety factor for a straight circular cylinder is given by $q = kr B_z / B_\theta$.

Physical mechanism: The deformation of the plasma increases the magnetic field pressure in regions where the magnetic field curvature is convex to the plasma interior, whilst simultaneously decreasing the magnetic pressure contribution where the deformation is concave (see Figure 9.4). This causes the plasma to continue to move towards the concave part of the deformation, so increasing the displacement.

9.4.5   Generalized Pinch Instabilities

The analysis of 9.4.4 is restricted to cylindrical plasmas with circular cross-section. The stability criteria can be extended to non-circular, non-cylindrical geometry using a generalised technique known as the *energy principle.*

The criteria for stability can be expressed in terms of the safety factor, q, defined by

$$q(r) = \frac{d\psi_{tor}}{d\psi_{pol}} \tag{9.64}$$

where ψ_{tor} and ψ_{pol} are the toroidal and poloidal magnetic fluxes, respectively, in toroidal geometry. The safety factor is a measure of the shear in the plasma magnetic field, and can be thought of as the ratio of the number of times a magnetic field line winds itself around the long way round a torus (the toroidal direction) to the number of times the same field line winds itself around the short way round a torus (the poloidal direction). Different geometries lead to different explicit formulae for q.

9.4.5.1 Energy Principle
The fundamental principle is that energy is conserved, so that the total of the kinetic and potential energies is always a constant. Hence any perturbation that decreases the potential energy must lead to a corresponding increase in the kinetic energy, that is, must produce an increased velocity perturbation. This indicates that the perturbation is linearly unstable. An excellent discussion of the subtleties in this theory is presented in [10]. The energy principle identifies linear instabilities, but does not directly predict growth rates for those instabilities.

Defining the plasma displacement $\boldsymbol{\xi}$ in terms of the perturbed plasma velocity $\boldsymbol{u}$,

$$\boldsymbol{\xi}(\boldsymbol{r}, t) = \int_0^t \boldsymbol{u}(\boldsymbol{r}, t') dt' \tag{9.65}$$

allows the linearised momentum equation for an ideal MHD plasma (7.86) to be written as

$$\rho_0 \frac{\partial^2 \boldsymbol{\xi}}{\partial t^2} = \boldsymbol{F}(\boldsymbol{\xi})$$

$$= \nabla(\boldsymbol{\xi} \cdot \nabla p_0 + \gamma p_0 \nabla \cdot \boldsymbol{\xi}) - \boldsymbol{B}_0 \times \{\nabla \times [\nabla \times (\boldsymbol{\xi} \times \boldsymbol{B}_0)]\}/\mu_0$$

$$+ (\nabla \times \boldsymbol{B}_0) \times [\nabla \times (\boldsymbol{\xi} \times \boldsymbol{B}_0)]/\mu_0 \qquad \text{(SI)} \tag{9.66}$$

$$= \nabla(\boldsymbol{\xi} \cdot \nabla p_0 + \gamma p_0 \nabla \cdot \boldsymbol{\xi}) - \boldsymbol{B}_0 \times \{\nabla \times [\nabla \times (\boldsymbol{\xi} \times \boldsymbol{B}_0)]\}/(4\pi)$$

$$+ (\nabla \times \boldsymbol{B}_0) \times [\nabla \times (\boldsymbol{\xi} \times \boldsymbol{B}_0)]/(4\pi) \qquad \text{(cgs)} \tag{9.67}$$

where quantities with a 0 subscript denote equilibrium values. The energy principle can then be expressed in terms of kinetic energy perturbations $\delta\mathcal{E}$ and potential energy perturbations δW as follows:

$$\frac{\partial}{\partial t}(\delta\mathcal{E} + \delta W) = \frac{\partial}{\partial t}\int_{plasma}\left(\tfrac{1}{2}\rho_0\dot{\xi}^2 + \tfrac{1}{2}\rho_0(\boldsymbol{\xi}\cdot\nabla)\boldsymbol{F}(\boldsymbol{\xi})\right)d\boldsymbol{r} = 0 \tag{9.68}$$

The potential energy contributions arising from the various physical effects can be summarised as

$$\delta W = \delta W_F + \delta W_V + \delta W_S \qquad (9.69)$$

where δW_F is the contribution arising from the fluid perturbations; δW_V comes from the vacuum magnetic field perturbation, and δW_S is the surface current contribution. Note that in plasmas with a *free-surface*, the analysis of instabilities involves all three contributions. However, *fixed boundary* plasmas do not take any surface or vacuum contribution into account.

These individual contributions can be quantified as follows:

fluid contributions:

$$\delta W_F = \tfrac{1}{2} \int_{plasma} (w_a + w_{fm} + w_{th} + w_k - w_i) \, \mathrm{d}\boldsymbol{r} \qquad (9.70)$$

where

$$w_a = \frac{B_\perp^2}{\mu_0} \qquad \text{Alfvén wave (SI)} \qquad (9.71)$$

$$= \frac{B_\perp^2}{4\pi} \qquad \text{Alfvén wave (cgs)} \qquad (9.72)$$

$$w_m = \frac{B_0^2}{\mu_0} \left| \nabla \cdot \boldsymbol{\xi} + \frac{\boldsymbol{\xi} \cdot \nabla \left(B_0^2 + 2\mu_0 p_0 \right)}{B_0^2} \right|^2 \qquad \text{Magnetosonic (SI)} \qquad (9.73)$$

$$= \frac{B_0^2}{4\pi} \left| \nabla \cdot \boldsymbol{\xi} + \frac{\boldsymbol{\xi} \cdot \nabla \left(B_0^2 + 8\pi p_0 \right)}{B_0^2} \right|^2 \qquad \text{Magnetosonic (cgs)} \qquad (9.74)$$

$$w_{th} = \gamma p_0 |\nabla \cdot \boldsymbol{\xi}|^2 \qquad \text{Acoustic} \qquad (9.75)$$

$$w_k = \frac{\boldsymbol{J}_0 \cdot \boldsymbol{B}_0}{B_0^2} (\boldsymbol{B}_0 \times \boldsymbol{\xi}) \cdot \boldsymbol{B} \qquad \text{Kink} \qquad (9.76)$$

$$w_i = 2(\boldsymbol{\xi} \cdot \nabla p_0)\boldsymbol{\xi} \cdot \boldsymbol{\kappa} \qquad \text{Interchange} \qquad (9.77)$$

and where $\boldsymbol{\kappa}$ is the magnetic curvature:

$$\boldsymbol{\kappa} = (\boldsymbol{B}_0 \cdot \nabla \boldsymbol{B}_0)/B_0^2 \qquad (9.78)$$

Note that in (9.71-9.77), subscript 0 denotes an equilibrium quantity, and $\perp$, $\parallel$ refer to components perpendicular and parallel to the direction of the equilibrium magnetic field $\boldsymbol{B}_0$. It is clear that only the kink w_k and interchange w_i terms can destabilise the plasma, since all other terms are positive definite.

vacuum contribution:

$$\delta W_V = \tfrac{1}{2} \int_{vacuum} |\boldsymbol{B}|^2/\mu_0 \mathrm{d}\boldsymbol{r} \quad \text{(SI)} \tag{9.79}$$

$$= \tfrac{1}{2} \int_{vacuum} |\boldsymbol{B}|^2/4\pi \mathrm{d}\boldsymbol{r} \quad \text{(cgs)} \tag{9.80}$$

quantifies the contribution from the vacuum magnetic field perturbation.

surface contribution:

$$\delta W_S = \int_{surface} (\hat{\mathbf{n}} \cdot \boldsymbol{\xi})^2 \nabla \big[\!| \left(p_0 + B_0^2/(2\mu_0)\right) |\!\big] \cdot \mathrm{d}\boldsymbol{S} \quad \text{(SI)} \tag{9.81}$$

$$= \int_{surface} (\hat{\mathbf{n}} \cdot \boldsymbol{\xi})^2 \nabla \big[\!| \left(p_0 + B_0^2/(8\pi)\right) |\!\big] \cdot \mathrm{d}\boldsymbol{S} \quad \text{(cgs)} \tag{9.82}$$

where the notation $[\!| \cdots |\!]$ signifies the jump at the boundary of the quantities enclosed by the brackets. Hence if there is no surface current present, (9.81) vanishes.

9.4.5.2 Suydam Criterion
When applied to straight cylinders of circular cross-section, the energy principle shows that stability demands

$$\left(\frac{q'}{q}\right)^2 + \frac{8\mu_0 p'}{rB_z^2} > 0 \quad \text{(SI)} \tag{9.83}$$

$$\left(\frac{q'}{q}\right)^2 + \frac{32\pi p'}{rB_z^2} > 0 \quad \text{(cgs)} \tag{9.84}$$

where the safety factor q in this context is given by

$$q(r) = kr\frac{B_z(r)}{B_\theta(r)} \tag{9.85}$$

9.4.5.3 Mercier Criterion
The energy principle analysis admits a more general stability description applicable to non-cylindrical plasmas, extending the Suydam criterion (9.83). For toroidal plasmas, with major radius R and minor radius r, the Mercier criterion governs the plasma stability for $r \ll R$, and $\beta \ll 1$:

$$\left(\frac{q'(r)}{q(r)}\right)^2 + \frac{8\mu_0}{r}\frac{p'(r)}{B_\varphi^2}(1 - q(r)^2) > 0 \quad \text{(SI)} \tag{9.86}$$

$$\left(\frac{q'(r)}{q(r)}\right)^2 + \frac{32\pi}{r}\frac{p'(r)}{B_\varphi^2}(1 - q(r)^2) > 0 \quad \text{(cgs)} \tag{9.87}$$

where $\boldsymbol{B}_\varphi$ is the toroidal magnetic field component, $\boldsymbol{B}_\theta$ is the poloidal magnetic field component, and the safety factor q is defined in this context as

$$q(r) = \frac{rB_\varphi(r)}{RB_\theta(r)} \qquad (9.88)$$

Note that the pressure and magnetic field depend only on r.

9.4.6 Resistive Drift Wave Instability

A semi-infinite plasma with a plane vacuum interface has a uniform magnetic field $\boldsymbol{B}_0$ present everywhere. Assume a density gradient n_0' perpendicular to the interface, and perturbations of the interface going as $\exp[\mathrm{i}(\boldsymbol{k} \cdot \boldsymbol{r} - \omega t)]$. Then the interface wave is linearly unstable, satisfying the dispersion relation

$$\omega^2 + \mathrm{i}\sigma(\omega - k_\perp v_{De}) = 0 \qquad (9.89)$$

where

$$\sigma = \frac{k_\perp^2}{k_\parallel^2}\omega_{ci}\omega_{ce}\tau_{ei} \qquad (9.90)$$

is the plasma conductivity. Here

$$v_{De} = \frac{n_0'}{n_0}\frac{k_B T_e}{eB_0} \qquad (9.91)$$

is the diamagnetic drift speed for electrons (6.46), τ_{ei} is the electron-ion collision time, and

$$k_\parallel = \boldsymbol{k} \cdot \boldsymbol{B}_0/B_0 \qquad (9.92)$$
$$\boldsymbol{k}_\perp = \boldsymbol{k} - k_\parallel \boldsymbol{B}_0/B_0 \qquad (9.93)$$

Note that $k_\perp \ll k_\parallel$, and $c_{th,i} \ll \omega/k_\parallel \ll c_{th,e}$.

Physical mechanism: Electrons drift diamagnetically along the interface, but lagging the fluid perturbation by virtue of the imperfect conductivity. This creates a space-charge density and a resultant electric field, which enhances the original interface distortion.

9.4.7 MHD Resistive Wall Instability

An ideal incompressible MHD flow parallel to a resistive wall is unstable if a critical flow speed is exceeded [95]. The wall has resistivity η and thickness δ, and is separated from the plasma by a vacuum region of distance d. The plasma has a uniform flow velocity $\boldsymbol{u}$ in the z-direction, parallel to the wall. This flow is unstable to perturbations of the form $\exp \mathrm{i}(kz - \omega t)$ if

$$|\boldsymbol{u}| > \sqrt{2}\,c_a \qquad (9.94)$$

with the frequency and growth rate given by

$$\omega \approx \frac{u^2 - 2c_a^2}{8c_w^2 + c_a^2 \exp(-4kd)} \left(\pm 2\sqrt{2}\, \frac{c_w}{c_a} + \mathrm{i}\exp(-2kd) \right) kc_w \qquad (9.95)$$

where

$$c_w = \frac{\eta}{\mu_0 \delta} \qquad \text{(SI)} \qquad (9.96)$$

$$= \frac{\eta c^2}{4\pi\delta} \qquad \text{(cgs)} \qquad (9.97)$$

In the low resistivity limit, the growth rate γ_g may be approximated as

$$\gamma_g = \mathrm{Im}(\omega) \approx \frac{u^2 - 2c_a^2}{c_a^2} \exp(2kd) kc_w \qquad (9.98)$$

Physical mechanism: The presence of the imperfectly conducting wall distorts the magnetic field associated with the interface perturbation, producing a time lag in the field evolution. For flows exceeding the critical speed, a current sheet on the plasma surface is induced which destabilises the plasma pressure in the perturbation.

9.4.8   MHD Resistive Tearing Mode

A sheared magnetic field in an incompressible resistive MHD plasma is subject to a tearing mode instability, in which magnetic islands form around a critical layer when the stationary equilibrium is disturbed.

Taking the gradient scale length of the equilibrium magnetic field $\boldsymbol{B}_0$ to be $\mathcal{L}$, the critical layer width to be $\epsilon\mathcal{L}$, the wavenumber of the perturbation to be k, and the plasma constant mass density and constant resistivity to be ρ and η respectively , then the growth rate γ of the instability can be estimated empirically to be [72]

$$\gamma \sim \alpha^{-2/5} \tau_A^{-2/5} \tau_R^{-3/5} \qquad (9.99)$$

where $\alpha = k\mathcal{L} > 1$, and τ_A, τ_R are the Alfvén transit time, and resistive diffusion time, respectively, defined in (2.13) and (2.15). The width of the critical layer is estimated to be

$$\epsilon\mathcal{L} \sim \left(\frac{\gamma\rho\eta c^2}{k^2 {B_0'}^2} \right)^{1/4} \qquad \text{(SI)} \qquad (9.100)$$

$$\sim \left(\frac{\gamma\rho\eta}{k^2 {B_0'}^2} \right)^{1/4} \qquad \text{(SI)} \qquad (9.101)$$

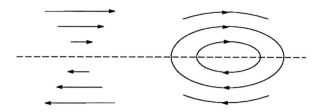

Fig. 9.5 Diagram showing the evolution of the tearing mode instability. The initial sheared magnetic field equilibrium is shown on the left, and the consequent magnetic island shown on the right.

where B_0' denotes the derivative of the equilibrium magnetic field.

A more detailed treatment extended to toroidal geometry yields [10]

$$\gamma \approx 0.55\tau_A^{-2/5}\tau_R^{-3/5}\alpha^{-4/5}\left(n\frac{\mathcal{L}^2 q'}{qR}\right) \tag{9.102}$$

where n is the toroidal mode number, R is the toroidal radius, and q is the safety factor (9.64); q' denotes the derivative of q with respect to minor radius.

The maximum growth rate γ_M can be estimated as [37, 89]

$$\gamma_M \sim 0.63S^{1/2}\tau_R^{-1} = 0.63(\tau_A\tau_R)^{-1/2} \tag{9.103}$$

which occurs at wavenumbers given by

$$\alpha \sim S^{-1/4} \tag{9.104}$$

Physical mechanism: The plasma in the critical layer ceases to be 'frozen' into the magnetic field, and becomes detached from it. The currents so generated create an additional magnetic field which permits the equilibrium field to reconfigure into an island structure (Figure 9.5, the latter state having lower energy than the initial one, and so representing an unstable process. Note that this is a long wavelength instability ($\alpha > 1$) of a stationary, incompressible resistive MHD plasma, with constant resistivity and mass density.

9.4.9   Streaming Instability

This instability is also known as the Buneman instability, the beam-plasma instability, and the two-stream instability.

The plasma is cold, unmagnetised and unbounded. Each plasma species has a constant equilibrium drift velocity $\boldsymbol{u}_{0s}$ with respect to some reference frame. The dispersion relation for small amplitude perturbations of this equilibrium, with frequency ω and wavevector $\boldsymbol{k}$ is

$$\sum_s \frac{\omega_{ps}^2}{(\omega - \boldsymbol{k} \cdot \boldsymbol{u}_{0s})^2} = 1 \qquad (9.105)$$

For the specific case of stationary ions (that is, observing from the ion rest-frame) (9.105) simplifies to

$$(\omega - \boldsymbol{k} \cdot \boldsymbol{u}_{0i})^2 (\omega^2 - \omega_{pi}^2) = \omega^2 \omega_{pe}^2 \qquad (9.106)$$

For low frequencies $\omega < \omega_{pi}$ (9.106) admits complex roots.

Physical mechanism: The perturbed motion of one stream of charged particles induces a mirror charge distribution in the other stream; for critical wavelengths, these charge concentrations interact sympathetically to enhance the amplitude of the wave, leading to unbounded growth of the perturbation.

9.5   KINETIC INSTABILITIES

9.5.1   Bump-in-Tail Instability

For an unmagnetized plasma, the general dispersion relation (7.137) has complex roots if the distribution function has two or more maxima in velocity space. A simple case is where the plasma distribution function consists of the superposition of two Maxwellian distributions at different temperatures:

$$f_0 = \frac{n_1}{n_0} a_1 \exp(-b_1 v^2)$$

$$+ \frac{n_2}{n_0} a_2 \frac{1}{2} \delta(v_x) \delta(v_y) \left[\exp\{-b_2(v_z - v_0)^2\} + \exp\{-b_2(v_z + v_0)^2\} \right]$$

$$(9.107)$$

where $a_i = m_e^{3/2} (2\pi k_B T_i)^{-3/2}$, $b_i = m_e/(2k_B T_i)$, δ is the delta function, and n_i/n_0 is the relative fraction of the total plasma at temperature T_i. The limiting case $n_2/n_0 \ll n_1/n_0$, $v_0 \gg 2k_B T_1/m_e$ is referred to as the 'bump-in-tail' instability [50]; the choice of model distribution function in (9.107) is to simplify the analysis by virtue of its symmetry.

The maximum growth rate γ_m for the bump-in-tail instability is given by

$$\gamma_m = \left(\frac{\pi}{8}\right)^{1/2} \frac{\omega_{p1}^2}{k^3 \lambda_{D1}^3} \left[\frac{n_2 T_1}{n_1 T_2} k^3 \lambda_{D1}^3 \frac{m_e v_0^2}{k_B T_1} e^{-1/2} - \exp\left(-\frac{1}{2k^2 \lambda_{D1}^2} - \frac{3}{2}\right) \right]$$

$$(9.108)$$

where

$$\lambda_{d1}^2 = \frac{\epsilon_0 k_B T_1}{n_1 e^2} \quad \text{(SI)} \tag{9.109}$$

$$= \frac{k_B T_1}{4\pi n_1 e^2} \quad \text{(cgs)} \tag{9.110}$$

Physical mechanism: The bump-in-tail instability is really a special, limiting case of the two-stream instability in Section 9.4.9.

9.5.2   Electron Runaway

An electric field $\boldsymbol{E}$ applied to an electron plasma with stationary ions will produce continuously accelerated electrons if $E > E_D$ where [4, 41]

$$E_D = \frac{\nu_{ei} m_e u}{e} \tag{9.111}$$

where ν_{ei} is the electron-ion collision frequency, m_e is the electron mass, and $u \sim c_{th,e}$ is the speed at which runaway is triggered. Using the Rutherford scattering formula for a 90° defection (6.6) as the form of ν_{ei} yields

$$E_D = \frac{n_i q_i^2 e \ln \Lambda}{(4\pi\epsilon_0)^2 m_e c_{th,e}^2} \quad \text{(SI)} \tag{9.112}$$

$$= \frac{n_i q_i^2 e \ln \Lambda}{m_e c_{th,e}^2} \quad \text{(cgs)} \tag{9.113}$$

E_D is known as the Dreicer field. For $E > E_D$, Ohm's law $\boldsymbol{J} = \sigma\boldsymbol{E}$ ceases to be valid. Those electrons in the electron distribution function with initial velocities parallel to $\boldsymbol{E}_0$ will become runaways, even for $E_0 < E_D$. The fraction of such runaways is given by

$$\frac{n_r}{n_e} \approx \frac{1}{2\pi} \exp(-E_D/E_0) \tag{9.114}$$

for $E_0 \ll E_D$.

Physical mechanism: The collisional drag term is speed dependent, and so particles which exceed a critical velocity under the influence of an applied electric field are exposed to unbalanced acceleration. Ultimately any such stream of runaway electrons will become unstable by other mechanisms.

9.5.3   Ion-Acoustic Instability

For an isotropic kinetic plasma in which the electrons at temperature T_e drift with speed u_0 through the ions, the latter having temperature $T_i \ll T_e$, the dispersion relation for waves is

$$\omega_r^2 = \frac{kc_{ia}^2}{1 + k^2\lambda_{De}^2} \tag{9.115}$$

$$\omega_i = -|\omega_r| \left(\frac{\pi}{8}\right)^{1/2} (1 + k^2\lambda_{De}^2)^{-3/2} \times$$

$$\left\{ \left(\frac{T_e}{T_i}\right)^{3/2} \exp\left[-\frac{1}{2}\frac{T_e}{T_i}(1 + k^2\lambda_{De}^2)^{-1}\right] \right.$$

$$\left. + \left(\frac{m_e}{m_i}\right)^{1/2} \left(1 - \frac{u_0}{c_{ia}}(1 + k^2\lambda_{De}^2)^{1/2}\right) \right\} \tag{9.116}$$

where $c_{ia} = (k_B T_i/m_e)^{1/2}$ is the ion-acoustic wave speed, and ω_r, ω_i are the real and imaginary parts of the wave frequency.

When the electron drift speed u_0 is zero, (9.115) and (9.116) are identical with the results of Section 7.4.3. However, if

$$u_0 > \frac{c_{ia}}{(1 + k^2\lambda_{De}^2)^{1/2}} > \left(\frac{k_b T_i}{m_i}\right)^{1/2} = c_{th,i} \tag{9.117}$$

then $\omega_i > 0$, and the ion-acoustic waves grow, rather than decay. For the limiting case $T_e \gg T_i$ so that the damping term can be ignored,

$$\omega_i = \left(\frac{\pi}{8}\frac{m_e}{m_i}\right)^{1/2} k\frac{u_0 - |\omega_r/k|}{(1 + k^2\lambda_{De}^2)^{3/2}} \tag{9.118}$$

Physical mechanism: The ion-acoustic mode is normally heavily damped, unless $T_i \ll T_e$, in which case it is only slightly damped. However, the streaming of one plasma component past another can induce instability, and so the weak drift of hot electrons past stationary, cool ions can be such that any residual damping is more than compensated for by the streaming instability.

10
Mathematics

10.1   VECTOR ALGEBRA

SYMBOL	DEFINITION
a, b, c, d	arbitrary vectors

$$a \cdot (b \times c) = b \cdot (c \times a) = c \cdot (a \times b) \tag{10.1}$$

$$a \times (b \times c) = b(a \cdot c) - c(a \cdot b) \tag{10.2}$$

$$(a \times b) \cdot (c \times d) = (a \cdot c)(b \cdot d) - (a \cdot d)(b \cdot c) \tag{10.3}$$

$$(a \times b) \times (c \times d) = (a \times b \cdot d)c - (a \times b \cdot c)d \tag{10.4}$$

$$a \times (b \times (c \times d)) = (b \cdot d)(a \times c) - (b \cdot c)(a \times d) \tag{10.5}$$

$$(a \times b) \cdot [(b \times c) \times (c \times a)] = [a \cdot (b \times c)]^2 \tag{10.6}$$

$$(ab) \cdot (cd) = a(b \cdot c)d \tag{10.7}$$

$$(ab) \times c = a(b \times c) \tag{10.8}$$

$$(ab) : (cd) = (a \cdot c)(b \cdot d) \tag{10.9}$$

$$(ab - ba) \cdot c = (b \times a) \times c \tag{10.10}$$

The definition of the dyad ab is

$$ab = \begin{bmatrix} a_1 b_1 & a_1 b_2 & a_1 b_3 \\ a_2 b_1 & a_2 b_2 & a_2 b_3 \\ a_3 b_1 & a_3 b_2 & a_3 b_3 \end{bmatrix} \tag{10.11}$$

where $a = (a_1, a_2, a_3)$ and $b = (b_1, b_2, b_3)$. Consequently, $ab \cdot c = a(b \cdot c)$.

10.2   VECTOR CALCULUS

SYMBOL	DEFINITION
f, g	arbitrary scalar functions
u, v	arbitrary vector functions

$$\nabla(fg) = f\nabla g + g\nabla f \tag{10.12}$$

$$\nabla \cdot (fu) = f\nabla \cdot u + u \cdot \nabla f \tag{10.13}$$

$$\nabla \times (fu) = f\nabla \times u + u \times \nabla f \tag{10.14}$$

$$\nabla \cdot (u \times v) = v \cdot \nabla \times u - u \cdot \nabla \times v \tag{10.15}$$

$$\nabla \times (u \times v) = u(\nabla \cdot v) - v(\nabla \cdot u) + (v \cdot \nabla)u - (u \cdot \nabla)v \tag{10.16}$$

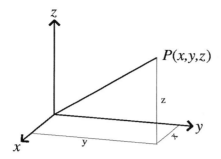

Fig. 10.1 The point P in the Cartesian co-ordinate system.

$$\nabla(\boldsymbol{u} \cdot \boldsymbol{v}) = \boldsymbol{u} \times (\nabla \times \boldsymbol{v}) + \boldsymbol{v} \times (\nabla \times \boldsymbol{u})$$

$$+ (\boldsymbol{u} \cdot \nabla)\boldsymbol{v} + (\boldsymbol{v} \cdot \nabla)\boldsymbol{u} \tag{10.17}$$

$$\nabla \cdot (\boldsymbol{uv}) = (\nabla \cdot \boldsymbol{u})\boldsymbol{v} + (\boldsymbol{u} \cdot \nabla)\boldsymbol{v} \tag{10.18}$$

$$\nabla \cdot (\boldsymbol{vu} - \boldsymbol{uv}) = \nabla \times (\boldsymbol{u} \times \boldsymbol{v}) \tag{10.19}$$

$$\nabla \times \nabla f = 0 \tag{10.20}$$

$$\nabla \cdot (\nabla \times \boldsymbol{u}) = 0 \tag{10.21}$$

$$\nabla \times (\nabla \times \boldsymbol{u}) = \nabla(\nabla \cdot \boldsymbol{u}) - \nabla^2 \boldsymbol{u} \tag{10.22}$$

$$\nabla \cdot \nabla f = \nabla^2 f \tag{10.23}$$

$$\nabla^2(fg) = f\nabla^2 g + 2(\nabla f) \cdot (\nabla g) + g\nabla^2 f \tag{10.24}$$

$$\nabla^2(f\boldsymbol{u}) = (\nabla f)(\nabla \cdot \boldsymbol{u}) + f\nabla(\nabla \cdot \boldsymbol{u}) + (\nabla f) \times \nabla \times \boldsymbol{u}$$

$$+ (\boldsymbol{u} \cdot \nabla)(\nabla f) + (\nabla f \cdot \nabla)\boldsymbol{u} \tag{10.25}$$

10.2.1 Cartesian Co-ordinates

Using the standard Cartesian co-ordinates (see Figure 10.1) (x, y, z) with respective unit vectors $\hat{\boldsymbol{x}}$, $\hat{\boldsymbol{y}}$ and $\hat{\boldsymbol{z}}$, we have:

$$\boldsymbol{u} = \hat{\boldsymbol{x}}u_x + \hat{\boldsymbol{y}}u_y + \hat{\boldsymbol{z}}u_z \tag{10.26}$$

$$\nabla \cdot \boldsymbol{u} = \frac{\partial u_x}{\partial x} + \frac{\partial u_y}{\partial y} + \frac{\partial u_z}{\partial z} \tag{10.27}$$

$$\nabla f = \hat{\mathbf{x}} \frac{\partial f}{\partial x} + \hat{\mathbf{y}} \frac{\partial f}{\partial y} + \hat{\mathbf{z}} \frac{\partial f}{\partial z} \qquad (10.28)$$

$$\nabla \times \mathbf{u} = \hat{\mathbf{x}} \left(\frac{\partial u_z}{\partial y} - \frac{\partial u_y}{\partial z} \right)$$

$$+ \hat{\mathbf{y}} \left(\frac{\partial u_x}{\partial z} - \frac{\partial u_z}{\partial x} \right)$$

$$+ \hat{\mathbf{z}} \left(\frac{\partial u_y}{\partial x} - \frac{\partial u_x}{\partial y} \right) \qquad (10.29)$$

$$\nabla^2 f = \frac{\partial^2 f}{\partial x^2} + \frac{\partial^2 f}{\partial y^2} + \frac{\partial^2 f}{\partial z^2} \qquad (10.30)$$

$$\nabla^2 \mathbf{u} = \hat{\mathbf{x}} \left(\nabla^2 u_x \right) + \hat{\mathbf{y}} \left(\nabla^2 u_y \right) + \hat{\mathbf{z}} \left(\nabla^2 u_z \right) \qquad (10.31)$$

$$(\mathbf{u} \cdot \nabla \mathbf{v}) = \hat{\mathbf{x}} \left(u_x \frac{\partial v_x}{\partial x} + u_y \frac{\partial v_x}{\partial y} + u_z \frac{\partial v_x}{\partial z} \right)$$

$$+ \hat{\mathbf{y}} \left(u_x \frac{\partial v_y}{\partial x} + u_y \frac{\partial v_y}{\partial y} + u_z \frac{\partial v_y}{\partial z} \right)$$

$$+ \hat{\mathbf{z}} \left(u_x \frac{\partial v_z}{\partial x} + u_y \frac{\partial v_z}{\partial y} + u_z \frac{\partial v_z}{\partial z} \right) \qquad (10.32)$$

$$\nabla (\nabla \cdot \mathbf{u}) = \hat{\mathbf{x}} \frac{\partial}{\partial x} \left(\frac{\partial u_x}{\partial x} + \frac{\partial u_y}{\partial y} + \frac{\partial u_z}{\partial z} \right)$$

$$+ \hat{\mathbf{y}} \frac{\partial}{\partial y} \left(\frac{\partial u_x}{\partial x} + \frac{\partial u_y}{\partial y} + \frac{\partial u_z}{\partial z} \right)$$

$$+ \hat{\mathbf{z}} \frac{\partial}{\partial z} \left(\frac{\partial u_x}{\partial x} + \frac{\partial u_y}{\partial y} + \frac{\partial u_z}{\partial z} \right) \qquad (10.33)$$

10.2.2 Cylindrical Co-ordinates

Using the standard cylindrical co-ordinates (see Figure 10.2)(ρ, ϕ, z) with respective unit vectors $\hat{\boldsymbol{\rho}}, \hat{\boldsymbol{\phi}}, \hat{\mathbf{z}}$, the following relationships hold:

$$\mathbf{u} = \hat{\boldsymbol{\rho}} u_\rho + \hat{\boldsymbol{\phi}} u_\phi + \hat{\mathbf{z}} u_z \qquad (10.34)$$

$$\hat{\boldsymbol{\rho}} = \hat{\mathbf{x}} \cos \phi + \hat{\mathbf{y}} \sin \phi \qquad (10.35)$$

$$\hat{\boldsymbol{\phi}} = -\hat{\mathbf{x}} \sin \phi + \hat{\mathbf{y}} \cos \phi \qquad (10.36)$$

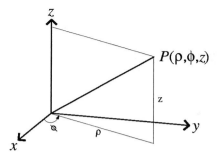

Fig. 10.2   The point P in the cylindrical co-ordinate system.

$$\mathrm{d}\boldsymbol{r} = \hat{\boldsymbol{\rho}}\,\mathrm{d}\rho + \hat{\boldsymbol{\phi}}\,\rho\mathrm{d}\phi + \hat{\boldsymbol{z}}\,\mathrm{d}z \tag{10.37}$$

$$\nabla \cdot \boldsymbol{u} = \frac{1}{\rho}\frac{\partial}{\partial \rho}(\rho u_\rho) + \frac{1}{\rho}\frac{\partial u_\phi}{\partial \phi} + \frac{\partial u_z}{\partial z} \tag{10.38}$$

$$\nabla f = \hat{\boldsymbol{\rho}}\,\frac{\partial f}{\partial \rho} + \hat{\boldsymbol{\phi}}\,\frac{1}{\rho}\frac{\partial f}{\partial \phi} + \hat{\boldsymbol{z}}\,\frac{\partial f}{\partial z} \tag{10.39}$$

$$\begin{aligned}
\nabla \times \boldsymbol{u} = \hat{\boldsymbol{\rho}}\,&\left(\frac{1}{\rho}\frac{\partial u_z}{\partial \phi} - \frac{\partial u_\phi}{\partial z}\right) \\
+ \hat{\boldsymbol{\phi}}\,&\left(\frac{\partial u_\rho}{\partial z} - \frac{\partial u_z}{\partial \rho}\right) \\
+ \hat{\boldsymbol{z}}\,&\left(\frac{1}{\rho}\frac{\partial}{\partial \rho}(\rho u_\phi) - \frac{1}{\rho}\frac{\partial u_\rho}{\partial \phi}\right)
\end{aligned} \tag{10.40}$$

$$\nabla^2 f = \frac{1}{\rho}\frac{\partial}{\partial \rho}\left(\rho\frac{\partial f}{\partial \rho}\right) + \frac{1}{\rho^2}\frac{\partial^2 f}{\partial \phi^2} + \frac{\partial^2 f}{\partial z^2} \tag{10.41}$$

$$\begin{aligned}
\nabla^2 \boldsymbol{u} = \hat{\boldsymbol{\rho}}\,&\left(\nabla^2 u_\rho - \frac{2}{\rho^2}\frac{\partial u_\phi}{\partial \phi} - \frac{u_\rho}{\rho^2}\right) \\
+ \hat{\boldsymbol{\phi}}\,&\left(\nabla^2 u_\phi + \frac{2}{\rho^2}\frac{\partial u_\rho}{\partial \phi} - \frac{u_\phi}{\rho^2}\right) \\
+ \hat{\boldsymbol{z}}\,&\left(\nabla^2 u_z\right)
\end{aligned} \tag{10.42}$$

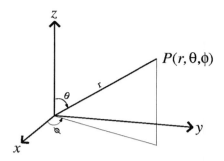

Fig. 10.3 The point P in the spherical co-ordinate system.

$$\boldsymbol{u} \cdot \boldsymbol{\nabla v} = \hat{\rho} \left(u_\rho \frac{\partial v_\rho}{\partial \rho} + \frac{u_\phi}{\rho} \frac{\partial v_\rho}{\partial \phi} + u_z \frac{\partial v_\rho}{\partial z} - \frac{u_\phi v_\phi}{\rho} \right)$$

$$+ \hat{\phi} \left(u_\rho \frac{\partial v_\phi}{\partial \rho} + \frac{u_\phi}{\rho} \frac{\partial v_\phi}{\partial \phi} + u_z \frac{\partial v_\phi}{\partial z} + \frac{u_\phi v_\rho}{\rho} \right)$$

$$+ \hat{z} \left(u_\rho \frac{\partial v_z}{\partial \rho} + \frac{u_\phi}{\rho} \frac{\partial v_z}{\partial \phi} + u_z \frac{\partial v_z}{\partial z} \right) \tag{10.43}$$

$$\nabla(\nabla \cdot \boldsymbol{u}) = \hat{\rho} \frac{\partial}{\partial \rho} \left(\frac{1}{\rho} \frac{\partial}{\partial \rho}(\rho u_\rho) + \frac{1}{\rho} \frac{\partial u_\phi}{\partial \phi} + \frac{\partial u_z}{\partial z} \right)$$

$$+ \hat{\phi} \frac{1}{\rho} \frac{\partial}{\partial \phi} \left(\frac{1}{\rho} \frac{\partial}{\partial \rho}(\rho u_\rho) + \frac{1}{\rho} \frac{\partial u_\phi}{\partial \phi} + \frac{\partial u_z}{\partial z} \right)$$

$$+ \hat{z} \frac{\partial}{\partial z} \left(\frac{1}{\rho} \frac{\partial}{\partial \rho}(\rho u_\rho) + \frac{1}{\rho} \frac{\partial u_\phi}{\partial \phi} + \frac{\partial u_z}{\partial z} \right) \tag{10.44}$$

10.2.3 Spherical Co-ordinates

Using the standard spherical co-ordinates (see Figure 10.3) (r, θ, ϕ) with respective unit vectors $\hat{\mathbf{r}}$, $\hat{\boldsymbol{\theta}}$ and $\hat{\boldsymbol{\phi}}$ we have:

$$\boldsymbol{u} = \hat{\mathbf{r}} \, u_r + \hat{\boldsymbol{\theta}} \, u_\theta + \hat{\boldsymbol{\phi}} \, u_\phi \tag{10.45}$$

$$\hat{\mathbf{r}} = \hat{\mathbf{x}} \sin(\theta)\cos(\phi) + \hat{\mathbf{y}} \sin(\theta)\sin(\phi) + \hat{\mathbf{z}} \cos(\theta) \tag{10.46}$$

$$\hat{\boldsymbol{\theta}} = \hat{\mathbf{x}} \cos(\theta)\cos(\phi) + \hat{\mathbf{y}} \cos(\theta)\sin(\phi) - \hat{\mathbf{z}} \sin(\theta) \tag{10.47}$$

$$\hat{\boldsymbol{\phi}} = -\hat{\mathbf{x}} \sin(\theta) + \hat{\mathbf{y}} \cos(\phi) \tag{10.48}$$

$$\mathrm{d}\boldsymbol{r} = \hat{\mathbf{r}}\,\mathrm{d}r + \hat{\boldsymbol{\theta}}\,r\mathrm{d}\theta + \hat{\boldsymbol{\phi}}\,r\sin\theta\mathrm{d}\phi \tag{10.49}$$

$$\nabla \cdot \boldsymbol{u} = \frac{1}{r^2}\frac{\partial}{\partial r}(r^2 u_r) + \frac{1}{r\sin\theta}\frac{\partial}{\partial\theta}(u_\theta\sin\theta) + \frac{1}{r\sin\theta}\frac{\partial u_\phi}{\partial\phi} \tag{10.50}$$

$$\nabla f = \hat{\mathbf{r}}\,\frac{\partial f}{\partial r} + \hat{\boldsymbol{\theta}}\,\frac{1}{r}\frac{\partial f}{\partial\theta} + \hat{\boldsymbol{\phi}}\,\frac{1}{r\sin\theta}\frac{\partial f}{\partial\phi} \tag{10.51}$$

$$\nabla \times \boldsymbol{u} = \hat{\mathbf{r}}\left(\frac{1}{r\sin\theta}\frac{\partial}{\partial\theta}(u_\phi\sin\theta) - \frac{1}{r\sin\theta}\frac{\partial u_\theta}{\partial\phi}\right)$$

$$+ \hat{\boldsymbol{\theta}}\left(\frac{1}{r\sin\theta}\frac{\partial u_r}{\partial\phi} - \frac{1}{r}\frac{\partial}{\partial r}(ru_\phi)\right)$$

$$+ \hat{\boldsymbol{\phi}}\left(\frac{1}{r}\frac{\partial}{\partial r}(ru_\theta) - \frac{1}{r}\frac{\partial u_r}{\partial\theta}\right) \tag{10.52}$$

$$\nabla^2 f = \frac{1}{r^2}\frac{\partial}{\partial r}\left(r^2\frac{\partial f}{\partial r}\right) + \frac{1}{r^2\sin\theta}\frac{\partial}{\partial\theta}\left(\sin\theta\frac{\partial f}{\partial\theta}\right)$$

$$+ \frac{1}{r^2\sin^2\theta}\frac{\partial^2 f}{\partial\phi^2} \tag{10.53}$$

$$\nabla^2 \boldsymbol{u} = \hat{\mathbf{r}}\left(\nabla^2 u_r - \frac{2u_r}{r^2} - \frac{2}{r^2}\frac{\partial u_\theta}{\partial\theta} - \frac{2u_\theta\cot\theta}{r^2} - \frac{2}{r^2\sin^2\theta}\frac{\partial u_\phi}{\partial\phi}\right)$$

$$+ \hat{\boldsymbol{\theta}}\left(\nabla^2 u_\theta + \frac{2}{r^2}\frac{\partial u_r}{\partial\theta} - \frac{u_\theta}{r^2\sin^2\theta} - \frac{2\cos\theta}{r^2\sin^2\theta}\frac{\partial u_\phi}{\partial\phi}\right)$$

$$+ \hat{\boldsymbol{\phi}}\left(\nabla^2 u_\phi - \frac{u_\phi}{r^2\sin^2\theta} + \frac{2}{r^2\sin\theta}\frac{\partial u_r}{\partial\phi} + \frac{2\cos\theta}{r^2\sin^2\theta}\frac{\partial u_\theta}{\partial\phi}\right) \tag{10.54}$$

$$\boldsymbol{u} \cdot \nabla \boldsymbol{v} = \hat{\mathbf{r}}\left(u_r\frac{\partial v_r}{\partial r} + \frac{u_\theta}{r}\frac{\partial u_r}{\partial\theta} + \frac{u_\phi}{r\sin\theta}\frac{\partial v_r}{\partial\phi} - \frac{u_\theta v_\theta + u_\phi v_\phi}{r}\right)$$

$$+ \hat{\boldsymbol{\theta}}\left(u_r\frac{\partial v_\theta}{\partial r} + \frac{u_\theta}{r}\frac{\partial v_\theta}{\partial\theta} + \frac{u_\phi}{r\sin\theta}\frac{\partial v_\theta}{\partial\phi} + \frac{u_\theta v_r - u_\phi v_\phi\cot\theta}{r}\right)$$

$$+ \hat{\boldsymbol{\phi}}\left(u_r\frac{\partial v_\phi}{\partial r} + \frac{u_\theta}{r}\frac{\partial u_\phi}{\partial\theta} + \frac{u_\phi}{r\sin\theta}\frac{\partial v_\phi}{\partial\phi} + \frac{u_\phi v_r + u_\phi v_\theta\cot\theta}{r}\right) \tag{10.55}$$

$$\nabla(\nabla \cdot \boldsymbol{u}) = \hat{\mathbf{r}} \frac{\partial}{\partial r} \left(\frac{1}{r^2} \frac{\partial}{\partial r}(r^2 u_r) + \frac{1}{r \sin\theta} \frac{\partial}{\partial \theta}(u_\theta \sin\theta) + \frac{1}{r \sin\theta} \frac{\partial u_\phi}{\partial \phi} \right)$$

$$+ \hat{\boldsymbol{\theta}} \frac{1}{r} \frac{\partial}{\partial \theta} \left(\frac{1}{r^2} \frac{\partial}{\partial r}(r^2 u_r) + \frac{1}{r \sin\theta} \frac{\partial}{\partial \theta}(u_\theta \sin\theta) + \frac{1}{r \sin\theta} \frac{\partial u_\phi}{\partial \phi} \right)$$

$$+ \hat{\boldsymbol{\phi}} \frac{1}{r \sin\theta} \frac{\partial}{\partial \phi} \left(\frac{1}{r^2} \frac{\partial}{\partial r}(r^2 u_r) + \frac{1}{r \sin\theta} \frac{\partial}{\partial \theta}(u_\theta \sin\theta) + \frac{1}{r \sin\theta} \frac{\partial u_\phi}{\partial \phi} \right)$$

$$(10.56)$$

10.3   INTEGRAL THEOREMS

SYMBOL	DEFINITION
C	closed curve forming the boundary of the surface S
S	regular 2-sided open surface with boundary C
$\mathrm{d}\boldsymbol{r}$	line element along C
$\hat{\mathbf{n}}$	unit outward normal to a surface
$\partial/\partial n$	directional derivative in the direction of $\hat{\mathbf{n}}$
$\mathrm{d}S$	scalar surface element
$\mathrm{d}\boldsymbol{S}$	$= \hat{\mathbf{n}}\mathrm{d}S$
Σ	regular closed surface containing volume V
V	volume contained by surface Σ
$\mathrm{d}\Sigma$	scalar surface element
$\mathrm{d}\boldsymbol{\Sigma}$	$= \hat{\mathbf{n}}\mathrm{d}\Sigma$
$\mathrm{d}V$	volume element
$\boldsymbol{u}$	arbitrary vector
f, g	arbitrary scalars

10.3.1   Stokes' Theorems

$$\oint_C f \, \mathrm{d}\boldsymbol{r} = \iint_S (\mathrm{d}\boldsymbol{S} \times \nabla) f \qquad (10.57)$$

$$\oint_C f \, \mathrm{d}g = -\oint_C g \, \mathrm{d}f = \iint_S \mathrm{d}\boldsymbol{S} \cdot (\nabla f \times \nabla g) \qquad (10.58)$$

$$\oint_C \boldsymbol{u} \cdot \mathrm{d}\boldsymbol{r} = \iint_S (\nabla \times \boldsymbol{u}) \cdot \mathrm{d}\boldsymbol{S} \qquad (10.59)$$

$$\oint_C \boldsymbol{u} \times \mathrm{d}\boldsymbol{r} = -\iint_S (\mathrm{d}\boldsymbol{S} \times \nabla) \times \boldsymbol{u} \tag{10.60}$$

10.3.2 Gauss' Theorems

$$\iint_\Sigma \boldsymbol{u} \cdot \mathrm{d}\boldsymbol{\Sigma} = \iiint_V \nabla \cdot \boldsymbol{u} \, \mathrm{d}V \tag{10.61}$$

$$\iint_\Sigma \boldsymbol{u} \times \mathrm{d}\boldsymbol{\Sigma} = -\iiint_V \nabla \times \boldsymbol{u} \, \mathrm{d}V \tag{10.62}$$

$$\iint_\Sigma f \mathrm{d}\boldsymbol{\Sigma} = \iiint_V \nabla f \, \mathrm{d}V \tag{10.63}$$

10.3.3 Green's Theorems

$$\iint_\Sigma f \frac{\partial g}{\partial n} \mathrm{d}\boldsymbol{\Sigma} = \iiint_V (f\nabla^2 g + \nabla f \cdot \nabla g) \, \mathrm{d}V \tag{10.64}$$

$$\iint_\Sigma \left(f \frac{\partial g}{\partial n} - g \frac{\partial f}{\partial n} \right) \mathrm{d}\boldsymbol{\Sigma} = \iiint_V (f\nabla^2 g - g\nabla^2 f) \, \mathrm{d}V \tag{10.65}$$

10.4 MATRICES

SYMBOL	DEFINITION
$\mathbf{M}$	arbitrary matrix of dimension $p \times q$
m_{ij}	i,jth entry of $\mathbf{M}$
$\mathbf{A}$	arbitrary square matrix of dimension $p \times p$
a_{ij}	i,jth entry of $\mathbf{A}$
$\mathcal{N}$	set of positive integers

The matrix $\mathbf{M}$ can be written in the form

$$\mathbf{M} = \begin{bmatrix} m_{ij} \end{bmatrix} \tag{10.66}$$

$$= \begin{bmatrix} m_{11} & m_{12} & \dots & m_{1q} \\ m_{21} & \dots & \dots & \dots \\ \dots & \dots & \dots & \dots \\ m_{p1} & \dots & \dots & m_{pq} \end{bmatrix} \tag{10.67}$$

where the indices $i = 1, \ldots, p$, and $j = 1, \ldots, q$, $p, q \in \mathcal{N}$. A square matrix is one where $p = q$.

10.4.1 Matrix Transpose

The transpose of $\mathbf{M}$ is obtained by exchanging rows with columns:

$$\mathbf{M}^{\mathrm{T}} = [m_{ji}]. \tag{10.68}$$

10.4.2 Complex Conjugate

The complex conjugate of a matrix is obtained by taking the complex conjugate of every entry:

$$\mathbf{M}^* = [m_{ij}^*] \tag{10.69}$$

10.4.3 Symmetric

$\mathbf{A}$ is symmetric if

$$\mathbf{A}^{\mathrm{T}} = \mathbf{A} \tag{10.70}$$

and is skew-symmetric if

$$\mathbf{A}^{\mathrm{T}} = -\mathbf{A}. \tag{10.71}$$

10.4.4 Orthogonal

$\mathbf{A}$ is orthogonal if

$$\mathbf{A}^{\mathrm{T}} = \mathbf{A}^{-1} \tag{10.72}$$

10.4.5 Nilpotent

$\mathbf{A}$ is nilpotent if

$$\mathbf{A}^k = 0 \tag{10.73}$$

for some $k \in \mathcal{N}$.

10.4.6 Idempotent

$\mathbf{A}$ is idempotent if

$$\mathbf{A}^2 = \mathbf{A} \tag{10.74}$$

10.4.7   Triangular

A is triangular if

$$a_{ij} = 0, \; j < i \quad \text{(upper triangular)} \qquad (10.75)$$
$$a_{ij} = 0, \; j > i \quad \text{(lower triangular)} \qquad (10.76)$$

A is *strictly* triangular if in addition to being triangular, $[a_{ii}] = 0, \; i = 1, \ldots, p$. All strictly triangular matrices are nilpotent.

The eigenvalues of a triangular matrix are the diagonal elements.

10.4.8   Trace

The trace of a square matrix is the sum of the entries in the main diagonal:

$$\operatorname{tr}(\mathbf{A}) = \sum_{i=1}^{p} a_{ii} \qquad (10.77)$$

10.4.9   Determinant and Inverse

The determinant of the general square matrix **A** is readily defined, albeit technical in construction. The interested reader is referred to [75] for further details. Instead we shall quote specific results for 2×2 and 3×3 matrices.

2×2 case

$$\det \mathbf{A} = |\mathbf{A}|$$

$$= \begin{vmatrix} a_{11} & a_{12} \\ a_{21} & a_{22} \end{vmatrix}$$

$$= a_{11}a_{22} - a_{12}a_{21} \qquad (10.78)$$

$$\mathbf{A}^{-1} = |\mathbf{A}|^{-1} \begin{bmatrix} a_{22} & -a_{21} \\ -a_{12} & a_{11} \end{bmatrix} \qquad (10.79)$$

3×3 case

$$\det \mathbf{A} = |\mathbf{A}|$$

$$= \begin{vmatrix} a_{11} & a_{12} & a_{13} \\ a_{21} & a_{22} & a_{23} \\ a_{31} & a_{32} & a_{33} \end{vmatrix}$$

$$= a_{11}a_{22}a_{33} - a_{11}a_{23}a_{32} - a_{12}a_{21}a_{33}$$
$$+ a_{12}a_{23}a_{31} + a_{13}a_{21}a_{32} - a_{13}a_{22}a_{31} \qquad (10.80)$$

$$\mathbf{A}^{-1} = |\mathbf{A}|^{-1} \times$$

$$\begin{bmatrix} a_{22}a_{33} - a_{23}a_{32} & -a_{12}a_{33} + a_{13}a_{32} & a_{12}a_{23} - a_{13}a_{22} \\ -a_{21}a_{33} + a_{23}a_{31} & a_{11}a_{33} - a_{13}a_{31} & -a_{11}a_{23} + a_{13}a_{21} \\ a_{21}a_{32} - a_{22}a_{31} & -a_{11}a_{32} + a_{12}a_{31} & a_{11}a_{22} - a_{12}a_{21} \end{bmatrix}$$

$$(10.81)$$

10.4.10 Partitioned Matrices

The array of elements belonging to the rows $i_1, i_2, \ldots, i_r$ and columns $j_1, j_2, \ldots, j_r$ of the matrix $\mathbf{M}$ constitutes a submatrix $\mathbf{M}_1$ of order $r \times s$.

Let $\mathbf{A}$ be the partitioned matrix

$$\mathbf{A} = \begin{bmatrix} \mathbf{A}_1 & \mathbf{A}_2 \\ \mathbf{A}_3 & \mathbf{A}_4 \end{bmatrix} \tag{10.82}$$

where the $\mathbf{A}_i$ are submatrices of the square matrix $\mathbf{A}$. Then if $\mathbf{A}_1$ and $\mathbf{A}_4$ posses inverses, then [3]

$$\mathbf{A}^{-1} = \begin{bmatrix} (\mathbf{A}_1 - \mathbf{A}_2\mathbf{A}_4^{-1}\mathbf{A}_3)^{-1} & (\mathbf{A}_2\mathbf{A}_4^{-1}\mathbf{A}_3 - \mathbf{A}_1)^{-1}\mathbf{A}_2\mathbf{A}_4^{-1} \\ (\mathbf{A}_3\mathbf{A}_1^{-1}\mathbf{A}_2 - \mathbf{A}_4)^{-1}\mathbf{A}_3\mathbf{A}_1^{-1} & (\mathbf{A}_4 - \mathbf{A}_3\mathbf{A}_1^{-1}\mathbf{A}_2)^{-1} \end{bmatrix}.$$

$$(10.83)$$

In particular, if all the $\mathbf{A}_i$ are invertible, then

$$\mathbf{A}^{-1} = \begin{bmatrix} (\mathbf{A}_1 - \mathbf{A}_2\mathbf{A}_4^{-1}\mathbf{A}_3)^{-1} & (\mathbf{A}_3 - \mathbf{A}_4\mathbf{A}_2^{-1}\mathbf{A}_1)^{-1} \\ (\mathbf{A}_2 - \mathbf{A}_1\mathbf{A}_3^{-1}\mathbf{A}_4)^{-1} & (\mathbf{A}_4 - \mathbf{A}_3\mathbf{A}_1^{-1}\mathbf{A}_2)^{-1} \end{bmatrix}. \tag{10.84}$$

10.4.11 Eigenvalues and Eigenvectors

The eigenvalues $\lambda_i, i = 1, \ldots, p$ of $\mathbf{A}$ are the roots of the characteristic polynomial defined by

$$|\lambda\mathbf{I} - \mathbf{A}| = 0. \tag{10.85}$$

Note that the determinant and trace of a square matrix can be expressed in terms of its eigenvalues:

$$\mathrm{tr}(\mathbf{A}) = \sum_{i=1}^{n} \lambda_i \tag{10.86}$$

$$|\mathbf{A}| = \prod_{i=1}^{n} \lambda_i \tag{10.87}$$

The eigenvectors $\xi_i, i = 1, \ldots, p$ associated with the eigenvalues of $\mathbf{A}$ are defined by

$$\mathbf{A}\xi_i = \lambda_i\xi_i \quad, i = 1, \ldots, p. \tag{10.88}$$

10.4.12 Hermitian Matrix

A Hermitian matrix is one for which the transpose is equal to its complex conjugate:

$$\mathbf{A}^{\mathrm{T}} = (\mathbf{A})^* . \qquad (10.89)$$

The matrix is skew-hermitian if

$$\mathbf{A}^{\mathrm{T}} = -(\mathbf{A})^* \qquad (10.90)$$

Clearly a real Hermitian matrix is a symmetric matrix. Note that Hermitian matrices have real eigenvalues.

10.4.13 Unitary Matrix

A unitary matrix has the property:

$$\mathbf{A}^{\mathrm{T}} = (\mathbf{A}^*)^{-1} \qquad (10.91)$$

Orthogonal matrices are then unitary ones with real entries.

10.5 EIGENFUNCTIONS OF THE CURL OPERATOR

SYMBOL	DEFINITION
$\boldsymbol{a}$	arbitrary vector
κ	parameter, with values ± 1
$\boldsymbol{Q}_\kappa$	unit vector corresponding to κ
$\boldsymbol{\Psi}_\kappa(\boldsymbol{r};\boldsymbol{a})$	eigenfunctions of curl operator

The eigenfunctions of the curl operator can be defined as follows [69]. Let the variable κ take the values 1,0,-1, and define the corresponding unit vectors $\boldsymbol{Q}_\kappa$ by

$$\boldsymbol{Q}_0(\boldsymbol{a}) = -\boldsymbol{a}/a, \qquad (10.92)$$

$$\begin{aligned}
\boldsymbol{Q}_\kappa(\boldsymbol{a}) = -\frac{\lambda}{\sqrt{2}} \times \Big\{ \\
+ \hat{\mathbf{x}} \left(\frac{a_x(a_x + \mathrm{i}\kappa a_y)}{a(a + a_z)} - 1 \right) \\
+ \hat{\mathbf{y}} \left(\frac{a_y(a_x + \mathrm{i}\kappa a_y)}{a(a + a_z)} - \mathrm{i}\lambda \right) \\
+ \hat{\mathbf{z}} \left(\frac{a_x + \mathrm{i}\kappa a_y}{a} \right) \Big\}
\end{aligned} \qquad (10.93)$$

Then the eigenfunctions $\boldsymbol{\Psi}_\kappa(\boldsymbol{r};\boldsymbol{a})$ of the curl operator can be defined in terms of the $\boldsymbol{Q}_\kappa(\boldsymbol{a})$:

$$\boldsymbol{\Psi}_\kappa(\boldsymbol{r};\boldsymbol{a}) = (2\pi)^{-3/2}\boldsymbol{Q}_\kappa(\boldsymbol{a})\exp(\mathrm{i}\boldsymbol{a}\cdot\boldsymbol{r}), \tag{10.94}$$

and possess the following properties:

$$\nabla\times\boldsymbol{\Psi}_\kappa(\boldsymbol{r};\boldsymbol{a}) = a\kappa\boldsymbol{\Psi}_\kappa(\boldsymbol{r};\boldsymbol{a}), \tag{10.95}$$

$$\nabla\cdot\boldsymbol{\Psi}_\kappa(\boldsymbol{r};\boldsymbol{a}) = 0 \text{ for } \kappa = \pm1, \tag{10.96}$$

$$\nabla\cdot\boldsymbol{\Psi}_0(\boldsymbol{r};\boldsymbol{a}) = -\mathrm{i}a(2\pi)^{-3/2}\exp(\mathrm{i}\boldsymbol{a}\cdot\boldsymbol{r}). \tag{10.97}$$

Equation (10.95) demonstrates that $\boldsymbol{\Psi}_\kappa(\boldsymbol{r};\boldsymbol{a})$ are the eigenfunctions of the curl operator, with corresponding eigenvalues $a\kappa$. Since the $\boldsymbol{Q}_\kappa(\boldsymbol{a})$ span the vector space, then any vector field can be decomposed, using the $\boldsymbol{\Psi}_\kappa(\boldsymbol{r};\boldsymbol{a})$ into 3 modes: one irrotational, and two of opposite normalised helicity. Further details about applications in electromagnetic theory and force-free magnetic fields are given in [63, 69].

10.6   WAVE SCATTERING

10.6.1   Simple Constant Barrier

second order scattering   The simplest case is a piece-wise continuous medium which supports waves of wavenumber k for $x < 0$ and $x > L$, but which has evanescent behaviour for $0 \le x \le L$. The differential equations are

$$\frac{\mathrm{d}^2y}{\mathrm{d}x^2} + k^2y = 0 \quad x < 0, \quad x > L \tag{10.98}$$

$$\frac{\mathrm{d}^2y}{\mathrm{d}x^2} - \kappa^2y = 0 \quad 0 \le x \le L \tag{10.99}$$

where κ characterises the evanescent behaviour inside the barrier. The form of the solution is then

$$y = e^{\mathrm{i}kx} + Re^{-\mathrm{i}kx} \quad x \le 0 \tag{10.100}$$

$$y = Ae^{\kappa x} + Be^{-\kappa x} \quad 0 \le x \le L \tag{10.101}$$

$$y = Te^{\mathrm{i}kx} \quad x \ge 0 \tag{10.102}$$

Solving for R, T, A and B yields

$$R = \frac{(\mu^2 - 1)(k^2 + \kappa^2)}{(\mu^2 - 1)(k^2 - \kappa^2) + 2\mathrm{i}k\kappa(\mu^2 + 1)} \tag{10.103}$$

$$T = -\frac{4\mathrm{i}k\kappa\mu e^{-\mathrm{i}kL}}{(k^2 - \kappa^2)(\mu^2 - 1) + 2\mathrm{i}k\kappa(\mu^2 + 1)} \tag{10.104}$$

$$A = 2k\frac{i\kappa - k}{(k^2 - \kappa^2)(\mu^2 - 1) + 2ik\kappa(\mu^2 + 1)} \tag{10.105}$$

$$B = 2k\mu^2\frac{i\kappa + k}{(k^2 - \kappa^2)(\mu^2 - 1) + 2ik\kappa(\mu^2 + 1)} \tag{10.106}$$

where $\mu = \exp(\kappa L)$. Energy conservation yields

$$|R|^2 + |T|^2 = 1, \tag{10.107}$$

a consequence of the fact that

$$\mathcal{W}(y, y^*) = \text{constant} \tag{10.108}$$

where $\mathcal{W}(u, v)$ is the Wronskian of the two functions $u(x)$ and $v(x)$, defined by

$$\mathcal{W}(u, v) = uv' - u'v, \tag{10.109}$$

fourth order scattering   Here there are 2 distinct waves, and so 4 possible wave solutions. The governing differential equation is

$$\frac{d^4u}{dx^4} + \alpha\frac{d^2u}{dx^2} + i\beta\frac{du}{dx} + \gamma u = 0 \tag{10.110}$$

where α, β and γ are real constants. The dispersion relation for waves of wavenumber k is

$$k^4 - \alpha k^2 - \beta k + \gamma = 0. \tag{10.111}$$

There are 3 invariant quantities $\mathcal{I}_1, \mathcal{I}_2$ and $\mathcal{I}_3$ associated with (10.110) [31]:

$$\mathcal{I}_1 = \mathcal{W}(u^{*\prime\prime}, u'') + i\beta u^{*\prime}u' - \gamma\mathcal{W}(u^*, u) \tag{10.112}$$

$$\mathcal{I}_2 = \mathcal{W}''(u^*, u) - 2\mathcal{W}(u^{*\prime}, u') + \alpha\mathcal{W}(u^*, u) + i\beta u^*u \tag{10.113}$$

$$\mathcal{I}_3 = u^{*\prime}u''' + u'u^{*\prime\prime\prime} - u^{*\prime\prime}u'' + \alpha u^{*\prime}u' + \gamma u^*u \tag{10.114}$$

where $\mathcal{W}(u, v)$ is the Wronskian of the two functions $u(x)$ and $v(x)$, defined by

$$\mathcal{W}(u, v) = uv' - u'v, \tag{10.115}$$

and where $u' = du/dx$.

Consider the particular case of a piecewise homogeneous medium having three distinct regions I, II and III, in which the physical properties of I and III are identical, but II is different. The definition of the regions, and the appropriate solutions to (10.110) in each can be taken as:

Region I: $-\infty < x < -L$ $u_I = \exp(ik_1x) + C_1\exp(-ik_1x)$

$$+ C_2 \exp(-ik_2 x) \tag{10.116}$$

Region II: $-L < x < L$ $\quad u_{II} = D_1 \exp(il_1 x) + D_2 \exp(-il_1 x)$
$$+ D_3 \exp(il_2 x) + D_4 \exp(-il_2 x) \tag{10.117}$$

Region III: $L < x < \infty$ $\quad u_{III} = F_1 \exp(ik_1 x) + F_2 \exp(ik_2 x) \tag{10.118}$

Then the the formula governing energy scattering between channels is:

$$|C_1|^2 + |F_1|^2 - \frac{k_2}{k_1}\rho\left(|C_2|^2 + |F_2|^2\right) = 1 \tag{10.119}$$

where

$$\rho = \frac{(k_2^2 - l_1^2)(k_2^2 - l_2^2)}{(k_1^2 - l_1^2)(k_1^2 - l_2^2)}. \tag{10.120}$$

10.6.2 Phase Integral Method

The phase-integral method, also known as WKB or JWKB, is a technique for determining the asymptotic solution to the ordinary differential equation [43]

$$\frac{d^2 u(x)}{dx^2} + h^2 q(x) u(x) = 0 \tag{10.121}$$

where h is a constant, and the wave potential $q(x)$ satisfies:

$q(x) \quad$ is continuous for all x; $\tag{10.122}$

$h^2 q(x) \to$ constant as $h \to \infty$ for fixed, arbitrary x; $\tag{10.123}$

$\left|\frac{dq/dx}{q}\right| \ll 1$ for all $|x| > x_c$; $\tag{10.124}$

$\left|\frac{d^2 q/dx^2}{q}\right| \ll 1$ for all $|x| > x_c$; $\tag{10.125}$

for some x_c.

The approximate, independent solutions $u_\pm$ to (10.121) for sufficiently large x take the form

$$u_\pm \sim \left[q^{-1/4} \exp\left(\pm ih \int^x q^{1/2} dx'\right)\right]\left[1 + O(h^{-2})\right], \tag{10.126}$$

where

$$O(h^{-2}) \sim \frac{5q'^2}{16q^2} - \frac{q''}{4q} \tag{10.127}$$

using the notation $q' = dq/dx$.

For the particular problem of waves encountering an overdense potential barrier, where the wave potential has zeros at $\pm a$ and takes the form

$$q(x) = x^2 - a^2, \quad (x, a \text{ real}) \tag{10.128}$$

then the energy reflection and transmission coefficients R and T take the form

$$|R|^2 = \frac{1}{1 + [a, -a]^2} \tag{10.129}$$

$$|T|^2 = \frac{[a, -a]^2}{1 + [a, -a]^2} \tag{10.130}$$

with

$$[a, b] = \exp(\mathrm{i}h \int_a^b q^{1/2} \mathrm{d}s). \tag{10.131}$$

The asymptotic solutions, valid for $|x| \gg a$ are given by (10.126) for left- and right-going waves. A more general treatment of coupled wave solutions can be found reference [42].

10.6.3 Mode Conversion

Approximate methods for calculating wave scattering in inhomogeneous media, without resorting to the complex and involved higher-order phase integral method or similar techniques, was developed based upon local dispersion relations and their properties. The basic idea is that the wavelength of a disturbance evolves continuously as it propagates in a spatially non-uniform plasma, reaching at least one critical position where there are two very simlar solutions to the local dispersion relation. Wave energy may then be distributed between otherwise distinct modes, in a process termed mode conversion.

For simple binary mode conversion [20] in an inhomogenous plasma, suppose the wave mode amplitudes are ϕ_1 and ϕ_2, and are governed by a local dispersion relation

$$(\omega - \omega_1)(\omega - \omega_2) = \eta \tag{10.132}$$

where $\omega_1(k, x)$ and $\omega_2(k, x)$ are the local frequencies of the two modes with corresponding amplitudes ϕ_i, k is the local 'wavenumber' in the direction of the inhomogeneity, and x is the independent variable. A wave of frequency ω_0 and wavenumber k_0 propagates through the inhomogeneous plasma, encountering the point x_0 at which $\omega_1(k_0, x_0) = \omega_2(k_0, x_0) = \omega_0$. Hence, near x_0,

$$k = k_0 + \delta, \tag{10.133}$$

$$x = x_0 + \xi, \tag{10.134}$$

$$\omega_1 = \omega_0 + a\delta + b\xi, \tag{10.135}$$

$$\omega_2 = \omega_0 + f\delta + g\xi \tag{10.136}$$

assuming Taylor expansions near x_0. The model equations to be solved are as follows, motivated by (10.132):

$$\frac{d\phi_1}{d\xi} - i(k_0 - \frac{b}{a}\xi)\phi_1 = i\left(\frac{\eta(x_0)}{af}\right)^{1/2}\phi_2,$$

$$\frac{d\phi_2}{d\xi} - i(k_0 - \frac{g}{f}\xi)\phi_1 = i\left(\frac{\eta(x_0)}{af}\right)^{1/2}\phi_1. \tag{10.137}$$

The solution with nonzero ϕ_1, ϕ_2 for $\xi < 0$, and only ϕ_1 nonzero for $\xi > 0$ yields an energy transmission coefficient

$$|T|^2 = \exp\left(-\frac{2\pi\eta(x_0)}{|ag - bf|}\right). \tag{10.138}$$

The energy 'reflection' coefficient is $|R|^2 = 1 - |T|^2$; however, the energy is deposited in ϕ_2, and so represents energy converted from the original mode.

Note that (10.138) is derived using the asymptotic expansion of the solution to (10.137), but that the latter is restricted to a small region around x_0 where (10.134)-(10.136) hold. Care must be taken to ensure that the local dispersion relation is a meaningful concept.

A generalized higher-order approach to mode conversion [90] concentrates on the solutions to a model comparison equation, termed the tunneling equation:

$$\frac{dy(z)}{dz^4} + \lambda^2 z \frac{d^2 y}{dz^2} + (\lambda^2 z + \gamma)y = 0 \tag{10.139}$$

The pseudo-dispersion relation for waves is derived from (10.139) in the form

$$k^4 - \lambda^2 z k^2 + \lambda z + \gamma \tag{10.140}$$

with approximate solutions

$$k_s \approx \pm(\lambda^2 z - 1)^{1/2} \tag{10.141}$$

$$k_f \approx \pm\left(1 + \frac{1 + \gamma}{\lambda^2 z}\right)^{1/2} \tag{10.142}$$

The asymptotic solution of (10.139) can be written as a superposition of these limiting modes, where f, s denote fast and slow solutions respectively. Power conservation can be written in the form

$$R^2 + T^2 + \rho C^2 = 1 \tag{10.143}$$

$$\rho = (1 - e^{-2\eta})^{-1} \qquad (10.144)$$

$$\eta = \pi \frac{1+\gamma}{2\lambda^2} \qquad (10.145)$$

for the case of an incident slow-mode yielding a reflected slow-mode component (R), a transmitted slow-mode component (T), and a mode-converted fast-mode (C) travelling in both directions. The precise proportion of converted mode and transmitted mode depend in detail on the actual physical problem, and general expressions are not easily presented in summary; see [90] for full details.

10.7   PLASMA DISPERSION FUNCTION

Linear wave analysis of a hot plasma with a Maxwellian equilibrium involves the plasma dispersion function, defined as [36]

$$\mathcal{Z}(\zeta) = \pi^{-\frac{1}{2}} \int_{-\infty}^{\infty} \frac{e^{-x^2}}{x-\zeta} dx \qquad (10.146)$$

or alternatively,

$$\mathcal{Z}(\zeta) = 2ie^{-\zeta^2} \int_{-\infty}^{i\zeta} e^{-t^2} dt \qquad (10.147)$$

The main properties of $\mathcal{Z}(\zeta)$ are as follows:

for all ζ:
$$\frac{d\mathcal{Z}}{d\zeta} = -2(1+\zeta\mathcal{Z}) \qquad (10.148)$$

zero argument:
$$\mathcal{Z}(0) = i\pi^{\frac{1}{2}} \qquad (10.149)$$

real argument:
$$\mathcal{Z}(x) = i\pi^{\frac{1}{2}} e^{-x^2} - 2xY(x) \qquad (10.150)$$

imaginary argument:
$$\mathcal{Z}(iy) = i\pi^{\frac{1}{2}} e^{y^2}[1 - \operatorname{erf}(y)] \qquad (10.151)$$

symmetry:
$$\mathcal{Z}(\zeta^*) = -[\mathcal{Z}(-\zeta)]^* \qquad (10.152)$$

for real x:
$$Y(x) = \frac{e^{-x^2}}{x} \int_0^x e^{t^2} dt \qquad (10.153)$$

Power series:
$$\mathcal{Z}(\zeta) = i\pi^{\frac{1}{2}} e^{-\zeta^2} - 2\zeta \left(1 - 2\zeta^2/3 + 4\zeta^4/15 + \cdots\right) \qquad (10.154)$$

Power series:
$$Y(x) = \tfrac{1}{2} x^{-2} \left(1 + 1/(2x^2) + 3/(4x^4) + \cdots\right) \qquad (10.155)$$

asymptotic expansion: $\mathcal{Z}(\zeta) = \mathcal{Z}(x + iy)$

$$\simeq i\pi^{\frac{1}{2}}\sigma e^{-\zeta^2} - \zeta^{-1}\left(1 + 1/(2\zeta^2) + 3/(4\zeta^4) + \cdots\right)$$
$$(10.156)$$

asymptotic expansion: $Y(x) \simeq \frac{1}{2}x^{-2}\left(1 + 1/(2x^2) + 3/(4x^4) + \cdots\right)$
$$(10.157)$$

where in (10.156),

$$\sigma = \begin{cases} 0 & y > 0 \\ 1 & y = 0 \\ 2 & y < 0 \end{cases} \qquad (10.158)$$

Appendix A
Guide to Notation

Symbol	Meaning	Ref
$\boldsymbol{A}$	magnetic vector potential	
b_0	critical impact parameter	(6.6)
b_s	ratio of thermal to wave mode energy	(7.162)
$\boldsymbol{B}$	magnetic flux density	
c	speed of light in vacuo	
c_a	Alfvén speed for the plasma	(2.24)
c_{as}	Alfvén speed for species s	(2.22)
$c_{th,s}$	sound speed for gas species s	(2.25)
c_{th}	gas sound speed	(2.25)
d	electrode separation	
d_s	planar sheath extent	
D_a	ambipolar diffusion coefficient	(3.42)
D_s	diffusion coefficient for species s	(3.32)
e	internal energy	(8.126)
E_D	Dreicer electric field	(9.111)
$\boldsymbol{E}$	electric field	
$\boldsymbol{E}_i$	incident electric field	
$\boldsymbol{E}_s$	scattered electric field	
f	distribution function	
f_0	equilibrium distribution function	
f_D	Druyvesteyn distribution function	(5.47)
f_M	Maxwell-Boltzmann distribution function	(5.9)
f_n	fractional electrostatic neutralisation	
F_M	f_M expressed as a distribution of speeds	(5.10)
$\mathcal{F}_\rho$	mass flux through a shock	(8.137)
g	energy distribution function	(5.6)
$\boldsymbol{g}$	acceleration due to gravity	
g_p	energy probability function	(5.7)
$\mathcal{G}$	Fokker-Planck potential	(5.36)
h	shock strength parameter	(8.136)
$\mathcal{H}$	Fokker-Planck potential	(5.37)
$\mathcal{H}_a$	Hartmann number	(2.38)
$\boldsymbol{H}$	magnetic intensity	
i_0	primary electron current at cathode	(3.56)
i_a	electron current at anode	(3.56)
I	current	
I_A	Alfvén current	(8.92)
I_{sc}	differential scattering cross-section	(2.29)
k_B	Boltzmann constant	
I_0	fundamental current in I_A	(8.95)

continued on next page

Symbol	Meaning	Ref
I_0	modified Bessel function of order 0	
I_n	modified Bessel function of order n	
$\boldsymbol{J}$	current density	
$\boldsymbol{J}_{ext}$	external current density	(5.20),(5.22)
J_m	Bessel function of 1st kind, order m	
J_i	ion current density	(3.9)
J_L	longitudinal invariant	(6.52)
$\boldsymbol{k}$	scattering wave vector	
$\boldsymbol{k}_i$	wave vector of incident electromagnetic wave	
$\boldsymbol{k}_s$	wave vector of scattered electromagnetic wave	
k_B	Boltzmann constant	
K	generalized beam perveance	(8.99)
K_m	modified Bessel function, order m	
$\mathbf{K}$	cold plasma dielectric tensor	(7.20)
m_r	reduced mass	(6.1)
m_s	mass of particle of species s	
$\mathcal{M}$	Mach number	(2.42)
n	refractive index of plasma, $= kc/\omega$	
n_c	cut-off density for an electron plasma	(7.56)
n_s	number density of particles of species s	
N_e	total number of electrons	(3.55)
N_{e0}	total number of electrons emitted at cathode	(3.55)
p	gas pressure	
P	power	
$\mathbf{P}$	pressure tensor	(5.3)
$\mathcal{P}$	total gas kinetic plus magnetic pressure	(7.128)
q	safety factor	(9.64)
q_s	charge carried by particle of species s	
$\boldsymbol{q}$	heat flux vector	(5.5)
Q	ionization rate	
r_e	classical electron radius	(4.72)
r_L	Larmor radius	(2.21)
R_m	magnetic Reynolds number	(2.43)
$\boldsymbol{r}$	position vector from origin to field point	(4.1)
$\boldsymbol{r}_0$	position vector from origin to source point	(4.1)
$\boldsymbol{R}$	position vector from source to field point	(4.1)
$\hat{\boldsymbol{R}}$	unit vector in $\boldsymbol{R}$ direction	
$\mathcal{R}_p$	pressure ratio across a shock front	(8.129)
$\mathcal{R}_\rho$	density ratio across a shock front	(8.130)
s	label defining species:i (ion), e (electron), n (neutral)	

continued on next page

SYMBOL	MEANING	REF
S	Lundquist number	(2.41)
T_g	gas (neutral) temperature	
T_p	plasma temperature	
T_s	temperature of gas of species s	
$\boldsymbol{u}$	fluid velocity	
u_r	relative speed	
$\boldsymbol{u}_D$	diamagnetic drift velocity	(6.44)
$\bar{\boldsymbol{u}}$	bulk or mean velocity	(5.2)
$\boldsymbol{u}$	bulk fluid plasma velocity	(7.81)
$\boldsymbol{u}_s$	velocity of species s	
u_0	ion speed at the plasma-sheath edge	(3.4)
u_i	ion speed in the sheath	(3.3)
V	voltage	
V_b	breakdown voltage	(3.73)
$V_{b,min}$	minimum breakdown voltage	(3.77)
$W_\parallel$	kinetic energy parallel to $\boldsymbol{B}$	(6.33)
$W_\perp$	kinetic energy perpendicular to $\boldsymbol{B}$	(6.33)
α	normalised wavenumber, $= k\lambda_D$	(4.64)
α_{T}	first Townsend ionization coefficient	(3.56)
$\boldsymbol{\beta}_v$	normalised particle velocity, $= \boldsymbol{v}/c$	
γ	polytropic index	
γ_g	growth rate	
γ_{T}	second Townsend ionization coefficient	(3.67)
Γ	gamma function	
Γ_c	fluid circulation	(8.5)
$\boldsymbol{\Gamma}_s$	flux of particles of species s	(3.28)
$\boldsymbol{\gamma}_v$	relativistic factor, $= (1 - \beta_v^2)^{-1/2}$	
δ	plasma skin depth	(2.20)
ϵ	ratio of photon energy to scatterer energy, $= \hbar\omega_i/(mc^2)$	
ε	energy density	(5.4)
ϵ_0	vacuum permittivity	
$\boldsymbol{\epsilon}$	hot plasma dielectric tensor	(7.152)
η	fluid plasma resistivity	
η_v	plasma viscosity	
λ_D	Debye length	(2.17)
$\lambda_{\mathrm{mfp},s}$	mean free path of species s	(2.19)
Λ	argument in Coulomb logarithm	(6.15)
μ_0	vacuum permeability	
μ_s	mobility of particle of species s	(2.34)
$\boldsymbol{\mu}_s$	mobility tensor for species s in a magnetised plasma	(2.36)

continued on next page

Symbol	Meaning	Ref
μ_{bs}	magnetic moment of a particle of species s	(2.33)
ν	non-specific collision frequency	
ν_B	Budker parameter	(8.87)
ν_c	electron-neutral collision frequency	
ν_{cs}	collision frequency of species s (in Hz)	
ν_{cs}	cyclotron frequency of species s (in Hz)	(2.9)
ν_{ps}	plasma frequency of species s (in Hz)	(2.3)
$\nu_{ss'}$	collision frequency for species s and s'	(2.12)
Π	polarization operator	(4.67)
ρ	mass density of single-fluid plasma	(7.80)
ρ_c	free charge density	
ρ_s	mass density of species s	
ρ_{ext}	external charge density	(5.19),(5.21)
σ_e	Thomson scattering cross-section for single electron	(4.69)
σ_j	square of sound over Alfvén speed on either side of shock	(8.132)
σ_{KN}	Klein-Nishina scattering cross-section	(4.100)
σ_{R}	Rutherford differential scattering cross-section	(6.2)
σ_{sc}	collision cross-section	(2.29)
τ_A	Alfvén transit time	(2.13)
τ_R	resistive diffusion time	(2.15)
ξ	energy loss factor	(5.40)
ω	frequency of electromagnetic wave	
ω_{cs}	circular cyclotron frequency of species s	(2.7)
ω_{ce0}	circular cyclotron frequency of rest electron	
ω_p	circular plasma frequency	(2.6)
ω_{ps}	circular plasma frequency of species s	(2.1)
$\boldsymbol{\omega}$	fluid vorticity	(8.6)
Ω	solid angle	

References

1. S Y Abdul-Rassak and E W Laing. *Journal of Plasma Physics*, 50:125–144, 1993.

2. M Abramowitz and I A Stegun. *Handbook of Mathematical Functions*. Dover, New York, USA, 1972.

3. A C Aitken. *Determinants and Matrices*. Oliver and Boyd, Edinburgh, UK, 1949.

4. A F Alexandrov, L S Bogdankevich, and A A Rukhadze. *Principles of Plasma Electrodynamics*. Springer-Verlag, Berlin, Germany, 1984.

5. H Alfvén. *Phys. Rev.*, 55:425–429, 1939.

6. H Alfvén. *Ark. f. Mat. Ast. Fys.*, 29A, 1958.

7. H Alfvén. *Reviews of Modern Physics*, 32:710–713, 1960.

8. H Alfvén and C-G Fälthammer. *Cosmical Electrodynamics*. Oxford University Press, London, UK, 2nd edition, 1963.

9. E S Aydil. Plasma etching. In G L Trigg, editor, *Encyclopedial of Applied Physics*. VCH Publishers, Wienheim, Germany, 1996.

10. G Bateman. *MHD Instabilities*. MIT Press, Cambridge, USA, 1978.

11. J Bazer and W B Ericson. *Astrophysical Journal*, 129:758–785, 1959.

12. G Bekefi and A H Barrett. *Electromagnetic Vibrations, Waves and Radiation*. MIT Press, Cambridge, USA, 1977.

13. A O Benz. *Plasma Astrophysics Kinetic Processes in Solar and Stellar Coronae*. Kluwer, Dordrecht, The Netherlands, 1993.

14. E Böhm-Vitense. *Introduction to Stellar Astrophysics*, volume 2: Stellar Atmospheres. Cambridge University Press, Cambridge, UK, 1969.

15. T J M Boyd and J J Sanderson. *Plasma Dynamics*. Nelson, London, UK, 1969.

16. S I Braginskii. *Soviet Physics JETP*, 6:358–369, 1958.

17. S I Braginskii. Transport processes in a plasma. In A D Leontovich, editor, *Reviews of Plasma Physics*, volume 1, pages 205–311. Consultants Bureau Enterprises, Inc, New York, USA, 1965.

18. S C Brown. *Introduction to Electrical Discharges in Gases*. John Wiley & Sons, New York, USA, 1966.

19. R A Cairns. *Plasma Physics*. Blackie, Glasgow, UK, 1985.

20. R A Cairns and C N Lashmore-Davies. *Phys. Fluids*, 26:1268–1274, 1983.

21. F F Cap. *Handbook on Plasma Instabilities*. Academic Press, New York, 1976.

22. S Chandrasekhar and L Woltjer. *Proc. Nat. Acad. Sciences*, 44:285–289, 1958.

23. F F Chen. *Introduction to Plasma Physics and Controlled Fusion*. Plenum Press, New York, USA, 1984.

24. P C Clemmow and J P Dougherty. *Electrodynamics of Particles and Plasmas*. Addison-Wesley, New York, 1969.

25. T G Cowling. *Month. Not. R. Astron. Soc.*, 94:39, 1934.

26. T G Cowling. *Magnetohydrodynamics*. Adam Hilger, Bristol, UK, 1976.

27. A A da Costa, D A Diver, and G A Stewart. *Astronomy & Astrophysics*, 366:129–137, 2000.

28. J M Dawson. *Phys. Fluids*, 5:445, 1962.

29. R O Dendy. *Plasma Dynamics*. Oxford University Press, Oxford, UK, 1990.

30. E A Desloge, S W Matthysse, and H Margenau. *Phys Rev*, 112:1437–1440, 1958.

31. D A Diver and E W Laing. *J Phys A: Math Gen*, 23:1699–1704, 1990.

32. A Egeland and N C Maynard. Ionospheric physics. In G L Trigg, editor, *Encyclopedia of Applied Physics*. VCH Publishers, Wienheim, Germany, 1996.

33. E M Epperlein and M G Haines. *Physics of Fluids*, 29:1029–1041, 1986.

34. U V Fahleson. *Phys. Fluids*, 4:123–127, 1961.

35. R N Franklin and J R Ockendon. *Journal of Plasma Physics*, 4:371–385, 1970.

36. B D Fried and S D Conte. *The Plasma Dispersion Function*. Academic Press, New York, USA, 1961.

37. H Furth, J Killeen, and M N Rosenbluth. *Phys. Fluids*, 6:459–484, 1963.

38. U Gebhardt and M Kiessling. *Phys. Fluids B*, 4:1689–1701, 1992.

39. V A Godyak and N Sternberg. *IEEE Transactions on Plasma Science*, 18:159–168, 1990.

40. D A Hammer and N Rostoker. *Phys. Fluids*, 13:1831, 1970.

41. R J Hastie. Introduction to plasma physics. In R D Gill, editor, *Plasma Physics and Nuclear Fusion Research*. Academic Press, London, UK, 1981.

42. J Heading. *Jour Res Nat Bur Stand D: Radio Propagation*, 65D:595–616, 1961.

43. J Heading. *An Introduction to Phase Integral Methods*. Methuen, London, 1962.

44. E H Holt and R E Haskell. *Foundations of Plasma Dynamics*. Macmillan, New York, USA, 1965.

45. A M Howatson. *An Introduction to Gas Discharges*. Pergamon Press, Oxford, UK, 1965.

46. I H Hutchinson. *Principles of Plasma Diagnostics*. Cambridge University Press, New York, USA, 1965.

47. J D Jackson. *Classical Electrodynamics*. John Wiley & Sons, New York, USA, 1962.

48. P C Kendall and R C Plumpton. *Magnetohydrodynamics with hydrodynamics*. Pergamon Press, London, UK, 1964.

49. E J Konopinski. *Electromagnetic Fields and Relativistic Particles*. McGraw-Hill, New York, USA, 1981.

50. N A Krall and A W Trivelpiece. *Principles of Plasma Physics*. McGraw-Hill, New York, USA, 1973.

51. H J Kunze. In W Lochte-Holtgreven, editor, *Plasma Diagnostics*. AIP Press, New York, USA, 1995.

52. E W Laing. *Plasma Physics*. Sussex University Press, London, UK, 1976.

53. J D Lawson. *Journal of Electronics and Control*, 3:587–594, 1957.

54. J D Lawson. *The Physics of Charged-Particle Beams*. Clarenden Press, Oxford, UK, 1977.

55. B Lehnert. *Physics of Fluids*, 10:2216–2225, 1967.

56. J S Lewis. *Physics and Chemistry of the Solar System*. Academic Press, London, UK, 1997.

57. D R Lide, editor. *CRC Handbook of Chemistry and Physics*. CRC Press, Florida, USA, 81st edition, 2000.

58. M A Lieberman and A J Lichtenberg. *Principles of Plasma Discharges and Materials Processing*. John Wiley & Sons, Inc, New York, USA, 1994.

59. F Llewellyn-Jones. *The Glow Discharge*. Methuen & Co Ltd, London, UK, 1966.

60. F Llewellyn-Jones. Theories of electrical breakdown of gases. In E E Kunhardt and L H Luessen, editors, *Electrical Breakdown and Discharges in Gases, Part A*, volume 86a of *NATO ASI Series B: Physics*. Plenum Press, New York, USA, 1983.

61. P Lorrain, D R Corson, and F Lorrain. *Electromagnetic Fields and Waves*. W H Freeman & Company, New York, USA, 1988.

62. Y M Lynn. *Physics of Fluids*, 9:314–335, 1966.

63. M A MacLeod. *Inverse Problems*, 11:1087–1096, 1995.

64. E W McDaniel. *Collision Phenomena in Ionized Gases*. John Wiley & Sons, New York, USA, 1964.

65. J F McKenzie and R K Varma. *Journal of Plasma Physics*, 25:491–497, 1981.

66. P J Mohr and B N Taylor. *Reviews of Modern Physics*, 72:351–495, 2000.

67. D C Montgomery and D A Tidman. *Plasma Kinetic Theory*. McGraw-Hill, New York, 1964.

68. W B Mori and T Katsouleas. *Physica Scripta*, T30:127–133, 1990.

69. H E Moses. *SIAM J Appl Math*, 21:114–144, 1971.

70. P Mulser. Laser light propagation and absorption. In M B Hooper, editor, *Laser-Plasma Interactions 3*, Proceedings of 29th Scottish Universities Summer School in Physics. SUSSP, Edinburgh, UK, 1986.

71. O Naito, H Yoshida, and T Matoba. *Physics of Fluids B*, 5:4256–4258, 1993.

72. S Ortolani and D D Schnack. *Magnetohydrodynamics of Plasma Relaxation*. World Scientific, Singapore, 1993.

73. E R Priest. *Solar Magnetohydrodynamics*. D Reidel, Dordrecht, Holland, 1984.

74. M A Raadu. *Physics Reports*, 178:27–97, 1989.

75. L Råde and B Westergren. *Mathematics Handbook for Science and Engineering*. Chartwell-Bratt Ltd, Kent, UK, 1995.

76. F H Read. *Electromagnetic Radiation*. John Wiley & Sons, Chichester, UK, 1980.

77. K-U Riemann. *J Phys D: Appl Phys*, 24:493–518, 1991.

78. M N Rosenbluth, W M MacDonald, and D L Judd. *Physical Review*, 107:1–6, 1957.

79. J R Roth. *Industrial Plasma Engineering*. Institute of Physics Publishing, Bristol, UK, 1995.

80. E E Salpeter. *Physical Review*, 120:1528–1535, 1960.

81. A C Selden. *Physics Letters*, 79A:405–406, 1980.

82. J Sheffield. *Plasma Scattering of Electromagnetic Radiation*. Academic Press, London, UK, 1975.

83. J A Shercliff. *Journal of Fluid Mechanics*, 9:481–505, 1960.

84. J A Shercliff. *A textbook of magnetohydrodynamics*. Pergamon Press, Oxford, UK, 1965.

85. I P Shkarofsky, T W Johnson, and M P Bachynski. *The Particle Kinetics of Plasmas*. Addison-Wesley, Reading, Massachusetts, 1966.

86. L J Spitzer, Jr. *Physics of Fully Ionized Gases*. Interscience Publishers, Inc., New York, USA, 1956.

87. G A Stewart and E W Laing. *Journal of Plasma Physics*, 47:295–319, 1992.

88. T H Stix. *Waves in Plasmas.* AIP, New York, USA, 1992.

89. P A Sturrock. *Plasma Physics.* Cambridge University Press, Cambridge, UK, 1992.

90. D G Swanson. *Plasma Waves.* Academic Press, San Diego, 1989.

91. J B Taylor. *Reviews of Modern Physics*, 58:741–763, 1986.

92. A W Trivelpiece and R W Gould. *Journal of Applied Physics*, 30:1784–1793, 1959.

93. R Y Tsien. *American Journal of Physics*, 40:46–56, 1972.

94. A L Ward. *Journal of Applied Physics*, 33:2789–2794, 1962.

95. J A Wesson. *Physics of Plasmas*, 11:3816–3819, 1998.

96. G Woan. *The Cambridge Handbook of Physics Formulas.* Cambridge University Press, Cambridge, UK, 2000.

97. L Woltjer. *Proc. Nat. Acad. Sciences*, 44:489–491, 1958.

98. L C Woods. *Principles of Magnetoplasma Dynamics.* Oxford University Press, Oxford, UK, 1987.

99. M V Zombeck. *Handbook of Space Astronomy and Astrophysics.* Cambridge University Press, Cambridge, UK, 1990.

Index